Mind and Morals

Mind and Morals

Essays on Cognitive Science and Ethics

edited by Larry May, Marilyn Friedman, and Andy Clark

A Bradford Book
The MIT Press
Cambridge, Massachusetts
London, England

This book was set in Palatino by Asco Trade Typesetting Ltd., Hong Kong and was printed and bound by in the United States of America.

Library of Congress Cataloging-in-Publication Data

Mind and morals: essays on cognitive science and ethics / edited by
  Larry May, Marilyn Friedman, and Andy Clark.
    p.   cm.
  "A Bradford book."
  Includes bibliographical references and index.
  ISBN 0-262-13313-X (alk. paper). — ISBN 0-262-63165-2
  (pbk.: alk. paper)
    1. Cognitive science—Moral and ethical aspects. 2. Cognitive science.
  3. Ethics. I. May, Larry. II. Friedman, Marilyn, 1945–   .
  III. Clark, Andy.
  BJ45.5.M56   1995
  170—dc20   95-22361
                                                            CIP

Contents

Acknowledgments

These chapters (with the exception of Peggy DesAutels's) were first delivered at a conference on Mind and Morals at Washington University in St. Louis in the spring term of 1994. Mark Rollins, Owen Flanagan, Marilyn Friedman, and Larry May came up with the idea for the conference during dinner at an APA Eastern Division Meeting in Boston in December 1992. Funding for the conference was quickly secured, with Roger Gibson's leadership, from Washington University's Philosophy, Neuroscience, and Psychology Program, the Department of Philosophy, and the Dean of the Faculty of Arts and Sciences. Several hundred faculty members and graduate students from North America and Europe attended the conference, the first of its kind to bring together philosophers interested in the intersection between ethics and cognitive science.

We are grateful to all of the faculty and students who made this conference, and this volume, such a success. We are especially grateful to Jason Clevenger for finding and transforming the art work used on the cover of the book; and to Henry Cribbs, John Feaster, Jo Forstrom, Mack Harrell, Adam Massie, Robert Stuffelbeam, and Josefa Toribio for their work on the index. We are also grateful to Teri Mendelsohn for first showing interest in this volume and shepherding us through the initial stages of production and to Sandra Minkkinen and the staff at the MIT Press for producing such a splendid volume.

Contributors

Michael Bratman is Professor of Philosophy and Senior Researcher at the Center for the Study of Language and Information at Stanford University. He is the author of *Intention, Plans,* and *Practical Reason* (1987) and articles on rational and intellectual agency.

Paul Churchland is Professor of Philosophy at the University of California San Diego. He is the author of *Matter and Consciousness* (1984), *A Neurocomputational Perspective* (1989), and *The Engine of Reason, The Seat of the Soul* (1995).

Andy Clark is Professor of Philosophy and Director of the Philosophy/Neuroscience/Psychology Program at Washington University in St. Louis. He is the author of *Microcognition* (1989) and *Associative Engines* (1993).

John Deigh teaches moral and political philosophy at Northwestern University. He is an associate editor and book review editor of *Ethics.*

Peggy DesAutels is a Postdoctoral Fellow at The Ethics Center at the University of South Florida, St. Petersburg. She has recently published in *Philosophical Psychology* and *The Journal of Social Philosophy.*

Owen Flanagan is Professor and Chair of Philosophy at Duke University. He is the author of *The Science of the Mind* (1991, second edition), *The Varieties of Moral Personality* (1991), and *Consciousness Revisited* (1992).

Marilyn Friedman is Associate Professor of Philosophy at Washington University in St. Louis. She is the author of *What are Friends For? Feminist Perspectives on Personal Relationships and Moral Theory* (1993) and coauthor of *Political Correctness: For and Against* (1995).

Alvin Goldman is Regents' Professor of Philosophy and Research Scientist in Cognitive Science at the University of Arizona. His books include *A Theory of Human Action* (1970), *Epistemology and Cognition* (1986), and *Philosophical Applications of Cognitive Science* (1993).

Robert M. Gordon is Professor of Philosophy at the University of Missouri at St. Louis. He is the author of *The Structure of the Emotions* (1987).

Virginia Held is Professor of Philosophy and Professor of Women's Stuides at Hunter College and the Graduate School of the City University of New York. Among her books are *The Public Interest and Individual Interests* (1970), *Rights and Goods* (1984), and *Feminist Morality* (1993).

Mark Johnson is Professor and Head of the Philosophy Department at the University of Oregon. He is the author of *Metaphors We Live By* (1980), *The Body in the Mind* (1987), and *Moral Imagination* (1993).

Susan Khin Zaw is Lecturer in Philosophy at the Open University in England. She has published papers on philosophy of mind, philosophy of biology, moral philosophy, practical ethics, and feminist history of ideas.

Helen E. Longino is Professor of Philosophy at Rice University. She is the author of *Science as Social Knowledge* (1990) and articles in philosophy of science and feminist theory.

Larry May is Professor of Philosophy at Washington University in St. Louis. He is the author of *The Morality of Groups* (1987), *Sharing Responsibility* (1992), and *The Socially Responsive Self* (1996).

Ruth Millikan is Professor of Philosophy at the University of Connecticut (fall terms) and the University of Michigan (spring terms). She is the author of *Language, Thought, and other Biological Categories* (1984), and *White Queen Psychology and Other Essays for Alice* (1993).

Naomi Scheman is Professor of Philosophy and Women's Studies at the University of Minnesota. She is the author of *Engenderings: Constructions of Knowledge, Authority and Privilege* (1993) as well as numerous articles in feminist epistemology.

James P. Sterba is Professor of Philosophy at the University of Notre Dame. His books include *The Demands of Justice* (1980), *How To Make People Just* (1988), and *Contemporary Social and Political Philosophy* (1994).

Chapter 1

Introduction

Larry May, Marilyn Friedman, and Andy Clark

At the beginning of the twentieth century, G. E. Moore decried what he called the naturalistic fallacy—the fallacy of conflating what people ought to do with what people actually do. Many philosophers were persuaded by Moore, with the result that ethics as a philosophical field was almost totally severed from psychology and sociology. By contrast, most of the contributors to this book think that moral philosophers should pay close attention to recent work in psychology, especially cognitive psychology. Perhaps their essays will help shape the agenda for the next generation of philosophers who work in ethics. These chapters dramatically illustrate how current work in cognitive science intersects with current work in moral philosophy, a connection that was palpably apparent in the 1994 conference on Mind and Morals at Washington University, from which this book emanates.

Work in psychology, like work in philosophy inspired by G. E. Moore, has mainly remained outside the moral domain. Of course, Piaget and Kohlberg did interesting cross-over work in moral psychology, but their contributions have been largely consigned to the margins of mainstream psychology, as is also true of large parts of social psychology that have touched on moral matters. But lately there has been increasing interest in moral reasoning on the part of cognitive and developmental psychologists. Moral reasoning is one of the most complex and difficult forms of reasoning, and any robust theory of reasoning is going to have to confront it eventually. It is one thing to show how a mind can calculate sums, but quite a different matter to show how a mind forms a judgment in a moral dilemma. The attempt to make cognitive science relevant to actual human reasoning has increased interest among cognitive scientists in moral philosophy. This interest will surely continue to grow in the new century.

The New Moral Naturalism

In *Principia Ethica*, one of the main foils of G. E. Moore's assault on ethical naturalism is John Stuart Mill's version of utilitarianism. Here is how

Moore puts it: "The whole object of Mill's book [*Utilitarianism*] is to help us discover *what we ought to do*; but, in fact, by attempting to define the meaning of this 'ought,' he has completely debarred himself from ever fulfilling that object: he has confined himself to telling us *what we do do*" (*Principia Ethica*, 73, our italics). It is one of the central premises of Moore's influential view of ethics that there is little or no relation between "what we ought to do" and "what we do do"—between ethics, on the one hand, and psychology, on the other hand.

Moore argued that anyone who blurred the distinction between the normative and the descriptive committed the "naturalistic fallacy" of confusing "two natural objects with one another, defining the one by the other" (*Principia Ethica*, 13). Ninety years have elapsed since Moore's challenge, and many, most notably William Frankena and John Searle, have shown that Moore was at best confused about the importance of this distinction (Frankena 1939; Searle 1964). There is surely a difference between saying that a practice such as incest exists in a certain society and saying that this practice is morally good. But to show that there is a difference is not at all to show that there is, and should be, no connection between these two claims.

Moore was a moral realist but not a moral naturalist. Moral realists believe in the possibility of establishing universal moral truth by reference to some kind of moral facts. These facts are normally thought to be grounded in the nonmoral facts about logic and the nature of reasoning. Moore appears to be a moral realist because he believed that "good" clearly named the property of having intrinsic value, and the truth of an assertion about whether something is good is evident to anyone who properly attends to it. But Moore was a nonstandard moral realist in that he believed that the term "good" named an existing property that was nonnatural but decidedly not nonmoral. That is, there were no facts by virtue of which moral truths could be established except for the moral facts ascertained by moral intuition. Psychology could play no role in establishing moral truth for Moore, because psychology concerned only nonmoral facts.

Moral naturalists are often confused with moral realists since both make reference to facts to establish the legitimacy of moral claims. But as the discussion of Moore shows, one can be a moral realist and not have any sympathy with moral naturalism. In addition, one can be a moral naturalist (such as Owen Flanagan or Mark Johnson in this book)[1] and yet not believe that there is universal moral truth. Most new-style moral naturalists wish to link morality and psychology but not in a way that reduces one to the other or makes the truth of moral claims depend on psychological facts. In this sense, most new moral naturalists have learned from Moore. They do not commit the naturalistic fallacy in any straightforward sense, for they neither define moral terms by natural ones nor base universal moral claims on empirical ones. But they do argue that an understanding of the way people actually make moral judgments is crucial for understanding what good moral judgments are.

Two forms of new moral naturalism are epitomized in some of the chapters that follow: normative ethical naturalism and metaethical naturalism. Normative ethical naturalism attempts to answer questions about what to do, value, or be using knowledge of how we actually behave, what we actually value, what sorts of persons we actually become. A radically naturalized ethics of this kind would hold that normative ethical questions were adequately and exhaustively answered by empirical knowledge of people's actual moral behavior and attitudes. This is the sort of view that Moore derided as committing the naturalistic fallacy.

Metaethical naturalism, by contrast, tries to answer questions about the methods we should use to *reach beliefs* about what to do, value, or be, and tries to answer these questions using empirical knowledge of the ways in which people actually reason morally. Contemporary moral naturalism is largely metaethical naturalism. Cognitive science, which focuses on human reasoning and thinking, has a unique contribution to make to ethics by way of metaethical naturalism.

In *Varieties of Moral Personality* (1991), Owen Flanagan argues that moral philosophers should adopt a principle requiring them to take psychology seriously. The principle is: "Make sure when constructing a moral theory or projecting a moral ideal that the character, decision processing, and behavior prescribed are possible, or are perceived to be possible, for creatures like us" (p. 32). This principle serves to mark off the moral domain but takes no position on how to assess ideals or norms that properly fall within that domain. So principles such as that proposed by Flanagan are examples of metaethical naturalism.

Various philosophers have made claims of a stronger sort for the relevance of psychology and other cognitive sciences to morality. Not only will cognitive science be able to explain how people actually reason morally; it may also be able to influence our moral choices. For instance, Paul Churchland and Peggy DesAutels believe that cognitive science may help us solve substantive moral quandaries in the same way that cognitive science helps us solve various visual problems. Churchland says that cognitive science will enable us to see that certain concepts of personhood are more appropriate than others, in much the same way that cognitive psychology can tell us which is the appropriate interpretation of an ambiguous visual object. And DesAutels says that cognitive science will help us see which is the better interpretation of a moral scenario, just as happens in gestalt shifts in visual perception. These views blur the distinction between metaethical naturalism and normative ethical naturalism.

Critics of these views are in ample supply. Virginia Held and James Sterba represent the loyal opposition in this book. Interestingly, most critics are sympathetic to many of the currents in cognitive science and moral psychology. They worry, however, that cognitive science will either attempt a hostile takeover of the field of ethics (Held), reducing all moral questions to psychological ones, or they worry about a very conservative tendency in moral psychology to place a premium on existing norms in particular societies regardless of how repressive those norms are (Sterba). These criticisms, though, seem to be mainly directed at normative ethical naturalism, not at that form of naturalism that merely asks that our conceptions of moral judgment conform to the reality of how people actually make those judgments.

The most often heard criticism is simply that when the new moral naturalists move into the domain of substantive moral norms and values, they fail to note that ethics attempts to tell us what people ought to do, not merely what they do. No investigation of brain states or evolutionary biology will convey to us what people should aspire to do or to be. This resurrects G. E. Moore's ghost, but perhaps now in a more palatable form. For example, even if cognitive scientists could describe the computational underpinnings of actual moral reasoning, this would fail to tell us anything at all about which moral judgments are better than others, for the best moral judgments may not yet have been thought by any brain.

In spite of such strong and important criticisms, all of the authors represented here find great value in a dialogue between cognitive scientists and moral philosophers. Ethics has always considered how theory realizes itself in practice; indeed ethics is often simply called practical philosophy. For this reason, if for no other, moral philosophers should be interested in any field of inquiry that gives a better picture of the practices that ethics comments upon. The fast-growing field of cognitive science is an obvious place to begin to look for new ideas about the practice of morality.

Cognitive Science and Its Potential Contribution to Ethics

Cognitive science brings to the ethical arena a developed interest in the kinds of internal states and resources that subserve thought and action. Our general understanding of the nature of these inner states and resources has undergone a rather dramatic shift in recent years. Many cognitive scientists have come to view the mind not as a repository of quasi-linguistic representations, rules, and recipes for action (the so-called classic cognitivist image; Fodor 1975, 1987) but as a medley of sophisticated pattern-completion devices exploiting very high-dimensional (supra-linguistic) modes of information storage. The interest of this alternative vision for ethics and moral philosophy lies in its implications for an empirically well-grounded conception of moral reasoning. Such implications (highlighted in the contributions of Churchland, Clark, and Flanagan) concern a variety of specific issues, such as the nature of moral knowledge, argument, and expertise (Churchland, Clark, Flanagan, and Sterba), the role of summary moral rules and maxims (Churchland, Clark), and the evolution and growth of moral understanding (Churchland, Flanagan, Deigh, and from a somewhat different angle, Millikan).

The roots of this alternative cognitive scientific vision of mind are many and various, but two important contributions are a body of cognitive psychological research that casts doubt on classical models of the structure of concepts and a body of computational work that develops a detailed and practicable alternative to classical "rule and symbol" invoking models of human cognition. The cognitive psychological work proposes what has become known as a prototype-based model of knowledge of concepts and categories, while the computational work proposes a connectionist (also known as neural network or parallel distributed processing) model of our basic internal computational resources. This computational vision, in turn, describes a perfect means by which to implement prototype-based knowledge in real brains. The two developments thus feed and support each other. No short introduction can hope to do justice to either, but it will help set the scene.

Grasping a concept, on the classical model, was depicted as involving knowing (perhaps unconsciously) a set of necessary and sufficient conditions for its application. Concepts thus defined by sets of necessary and sufficient conditions should have clear-cut conditions of application. Yet a body of cognitive psychological research (Rosch 1973; Smith and Medin 1981) shows instead that most concepts are judged to apply to instances to a greater or lesser extent—so-called typicality effects in which, say, a dog is judged to be a better pet than a dove, though both can fall under the concept. Such typicality effects are explained if we suppose that we organize our knowledge not around quasi-linguistic definitions but around stored knowledge of prototypical instances. This idea of prototypes as the pivotal mode of internal representation of knowledge is crucial to the reconception of moral cognition (see especially chapter 5 by Churchland and critical comments from Sterba and Flanagan) and requires expansion.

We should carefully distinguish exemplars, stereotypes, and prototypes. Exemplars are the concrete instances we encounter during training or learning. Stereotypes are the socially constructed images of "typical" exemplars of a concept or category (e.g., the stereotypic nurse). Prototypes (as used in most of the cognitive scientific literature) are the internally represented results of a process that extracts statistical central tendency information from the specific set of exemplars to which an individual system has been exposed. Statistical central tendency information is information concerning which features are most common to the exemplars of some class. One of the central tendencies of the "pet" class may be the feature "furry." The idea of a prototype is thus the idea of a body of features united as the most statistically common characteristics of the exemplars to which a system has been exposed. The prototype is a point or region in a high-dimensional space (one dimension for each represented feature), and the set of features that comprise it need not correspond to any actual exemplar or socially constructed stereotype. The prototypical saint may comprise a set of features never found together in any actual person. And there is no bound on the kinds of thing, process, or event for which we may command a prototype. Our knowledge about economic trends may be organized around economic prototypes (depression, recovery, etc.), and we may command social, legal, and moral prototypes of a wide variety of kinds (Churchland).

Connectionist computational models (McClelland, Rumelhart, and the PDP Research Group 1986, vols. 1, 2; Churchland 1989; Clark 1989) offer a neuroscientifically plausible general vision of how prototype-style knowledge may be represented in real brains. Such models store knowledge by gradually adapting a large web of simple processing units and connections. Units are connected by signal-carrying channels ("connections") that are weighted so as to increase or inhibit activity along them. A unit typically receives inputs in parallel from several other units along such channels. It sums and transforms such inputs according to a specific function and passes the resulting set of signals along some further set of outgoing parallel connections. Activity is thus propagated through the system in a way dictated by the particular pattern of the excitatory and inhibitory weights on the connections. It is these weights that are typically adapted during learning so that the overall system comes to implement a particular input-output function.

Such networks of units and weights are able to implement prototype-based knowledge structures. The networks are (typically) exposed to an extensive sample of exemplars—specific inputs with associated target outputs. An automatic learning algorithm gradually adapts the connection weights so as to improve the network's (initially quite hopeless) performance. Extended training results (if all goes well) in the network's extraction of the statistical central tendency of the example cases. More precisely, the network comes to use its representational resources (the units and weights) so as to create a systematically organized high-dimensional space. Inputs cause activity, which picks out points or regions in that space. Such points or regions encode specific feature complexes (prototypes) that proved salient during training. Novel inputs are assimilated to such familiar regions according to their degree of similarity to the training cases. In this way the system can produce sensible outputs for never-before-encountered inputs. (For a fuller treatment, see Clark 1989.)

Reasoning and inference are reconstructed within this connectionist paradigm as computationally continuous with basic abilities of pattern completion and pattern generalization. Moral failure and moral insight are thus explained as, in effect, rather subtle cases of perceptual failure and perceptual acuity. The good agent commands a set of developed moral prototypes that structure the way she processes inputs and hence, in a profound sense, the way she sees the world. (This theme is developed in the chapters by Churchland and DesAutels.)

Moreover, the kind of know-how thus encoded will often be, in a quite unmysterious way, supraverbal, for the internal representational spaces that house the mature agent's array of learned prototypes can easily assume gargantuan proportions. Even relatively small networks of units and weights command representational spaces of high dimensionality and great expressive power. To hope to condense the subtle moral expertise encoded by a biologically realistic neural network into a tractable set of summary principles or moral rules is in all probability a quite hopeless task. In this respect, moral expertise looks set to follow the pattern of expertise in general. Long volumes fail dismally to capture the knowledge of the expert chess player. The knowledge of the successful moral agent is probably no less complex, and no more amenable to brief linguaform summation (see Churchland, Flanagan, and Clark).

The nature of the internal representation of moral knowledge is a theme also taken up, with a somewhat different slant, by Ruth Millikan. Millikan's focus is on one particular—hypothesized—subspecies of internal representation. This subspecies, which she terms "pushmi-pullyu representations," is unusual in that it functions both to describe states of affairs and to prescribe courses of action. A being that represented its environment as a set of possibilities of motion and action would be exploiting such dual-purpose representations. Pushmi-pullyu representations, Millikan conjectures, may be more evolutionarily and developmentally basic than more familiar constructs such as detached beliefs and other so-called propositional attitudes. In the moral realm, pushmi-pullyu representations may occupy a crucial space between the purely descriptive and the out-and-out prescriptive. It is perhaps worth noting (although Millikan herself does not pursue this idea) that the basic representations learned by simple connectionist networks exhibit just such a dual character: they encode information about the world in a way that combines procedures of categorization with decisions about actions. Thus, the famous text-to-phoneme translation network NETtalk (Sejnowski and Rosenberg 1987) encodes knowledge about the grammatical grouping "vowels," but this knowledge is available only as part of a routine that uses the knowledge to yield specific behaviors (in this case, the production of coding for speech). (For discussion of this feature of basic connectionist encodings, see Clark and Karmiloff-Smith 1993; Clark 1993.)

The themes of moral know-how and moral reason are also taken up by Susan Khin Zaw in her critical discussion of classical models of moral rationality. Khin Zaw argues for a reconceptualization of moral rationality that places practical action and the invention of constraint-satisfying solutions at the heart of moral reason. The kinds of delicate and multiple-constraint-satisfying judgments and perceptions that this notion demands are, she speculates, the proper objects of further cognitive scientific research, research that can help us understand how practical reason operates and further illuminate its character.

Perspectives on internal representation have direct implications for a variety of more specific issues of interest to moral philosophy. Paul Churchland discusses the nature of moral argument, depicting moral disagreement as, for the most part, not disagreement over the truth or falsity of explicit moral rules but rather as a contest between alternative prototypes offered as "correct" avenues for the understanding of some state of affairs (for example, the contest between "nonviable fetus" and "person" prototypes as ways of understanding the moral problems posed by abortion.) (See chapter 13 by Sterba for a critical discussion of the putative difference between rule and prototype accounts.)

Churchland also raises issues concerning moral realism and moral progress. The same kinds of pressures (for successful, workable solutions) that incline biological brains to develop useful representations of the physical world hold equally, Churchland suggests, in the social and moral realms. Hence, our overall progress toward a better and more convergent scientific understanding will be matched by an increasingly progressive and convergent moral and social understanding. Owen Flanagan endorses Churchland's general image of the internal representational basis of moral reason but argues instead for a nonconvergent plurality of "good ways of living." This moral pluralism is justified, he claims, by the comparative locality (both temporal and spatial) of the specific conditions to which moral understanding responds. This localization, Flanagan believes, constitutes a deep disanalogy with the case of basic physics and invites instead a parallel between moral understanding and human ecology. What we need, Flanagan suggests, is a kind of moral ecology, that is, the study of what contributes to the well-being of human groups in specific ecological niches.

Flanagan also raises important questions concerning the role of society and interpersonal exchange in the development of moral knowledge. This theme is taken up by Clark, who stresses the role of collaborative reasoning in moral cognition. Practical moral reason, Clark suggests, does not always involve direct competition between alternative prototype-based understandings (compare Churchland). Instead it is often a matter of seeking a practicable course of action (see also Khin Zaw) that accommodates such multiple perspectives. Once the collaborative dimensions of moral cognition are fully recognized, Clark claims, it becomes possible to see summary moral rules and maxims in a new light. Individual moral knowledge may indeed be "supraverbal," as Churchland and others suggest, but the summary formulations in which linguaform moral debates trade may be best understood not as attempts to capture the full intricacies of individual moral know-how but rather as essential guides and signposts designed to promote the kinds of minimal mutual understanding necessary for the collaborative confrontation of practical moral issues. Underlying such collaborative efforts, Clark suggests, lie a variety of second-order communicative prototypes: ideas concerning what other agents already know, what they need to know to enter into collaborative activity, and how best to convey to them what they need to know. The internal representation of such second-order knowledge must be addressed, Clark argues, before connectionism can claim to illuminate fully issues concerning moral cognition.

Michael Bratman's chapter highlights a further capacity essential to sophisticated social coordination: the capacity to make plans and to (usually) stick to them. Attention to this important capacity, he argues, can help illuminate the nature of temptation and suggest strategies for resisting.

Other issues addressed from a cognitive scientific perspective include the character of moral expertise (Clark, Churchland, DesAutels, Flanagan), the nature and roots of moral pathology (Churchland, Deigh), the encoding of moral knowledge in specific areas of the brain (Churchland, Longino), the role of imagination and metaphorical understanding in moral reason (Johnson), the nature of early moral knowledge and moral education (Churchland, Deigh, Sterba, Millikan), and the role of simulation and empathy in moral thought. The last theme is placed at center stage in the chapters by Goldman, Gordon, and Deigh. Goldman argues that empathy and simulation could in principle underpin the legitimacy of the interpersonal utility comparisons required by certain theories within moral philosophy and economics, but that whether such capacities in fact do legitimize such comparisons turns out to be an empirical question open to cognitive scientific investigation. Gordon highlights the role of simulation in various morally sensitive situations (such as giving advice and judging actions) and argues that in such cases the simulation must be somewhat restricted to allow room for appraisal from the point of view of the judging-advising agent (that is, we must not simply try to replicate the target's moral and emotional profile). Deigh similarly locates empathy as a key factor in moral understanding and argues that this factor is especially impaired in psychopathic thought.

The Potential Contribution of Ethics to Cognitive Science

Johnson depicts the main role of moral theory as the "enrichment and cultivation of moral understanding." Empirical research in psychology and cognitive science is relevant, he argues, to the development of such understanding by providing facts about human motivation, learning, the nature of moral reason, the constraints on moral inference, and so on. Held argues, by contrast, that cognitive science cannot, in fact, offer a great deal to moral philosophy since the latter deals largely not in explanation (the province of cognitive science) but in recommendation. Understanding the causal mechanisms underlying judgments, Held writes, will not help us to gauge their justification as normative recommendations. Held thus critically discusses the work on empathy, prototypes, and the role of metaphor in moral thought as pursued in the chapters by Goldman, Churchland, and Johnson.

Cognitive science indeed appears to leave substantial room for both the autonomy of ethics and the potential for ethics to contribute to scientific studies of moral cognition. First, cognitive scientific theories about the nature of moral cognition seem to resolve few, if any, normative moral problems. Second, there seem to be various ways in which cognitive science research on moral cognition continues to depend on prior ethical concepts and assumptions. Third, there are normative moral issues raised by the very practices of cognitive science, which moral philosophers can helpfully address. We will discuss each of these points in turn.

First, cognitive scientific theories about the nature of moral cognition seem to resolve few, if any, normative moral problems. Even if cognitive science could provide a reliable account of how people had to (or could not) think morally, nevertheless such an account would not necessarily dictate unique normative conclusions to particular moral questions. As Held stresses, there is a residual gap between questions of what we ought to do, value, or be and questions of how we ought to arrive at conclusions about what to do, value, or be. Particular methods of moral reasoning might constrain moral reasoning in certain directions, but there is no reason to think that they determine specific beliefs as outcomes. So long as people using identical reflective strategies could plausibly reach different and even contradictory moral conclusions about the same issues, then cognitive scientific accounts of moral reasoning would not necessarily determine any matters of normative moral substance. Teachers of applied ethics have long observed that consequentialist and deontological ethical principles can each be used to defend as well as to criticize the same human behaviors, depending on what additional conceptual or empirical assumptions accompany the principles. There is no reason to think that this flexibility would not also characterize the exemplar-based models of moral reasoning now favored by cognitive science.

Cognitive science by itself will not enable us to decide, for example, whether the belief that capital punishment is morally permissible is cognitively superior to the belief that it is not, or whether the belief that wealthy nations have a moral responsibility to provide food aid to poor nations is cognitively superior to the belief that they do not. The debate over normative moral beliefs appears to exceed what a metaethical naturalism, bolstered by cognitive science, could tell us. In this regard, ethical problems as a whole would not be adequately or exhaustively resolved by the approach of contemporary ethical naturalism. The contemporary naturalistic turn in ethics thus has important limits. (This bodes well for the careers of moral philosophers!)

Second, there are various ways in which cognitive science research on moral cognition continues to depend on prior ethical concepts and assumptions. For one thing, cognitive scientists must differentiate moral cognition from nonmoral cognition. Cognitive scientists can make such distinctions by relying on their own folk psychological intuitions about what count as instances of moral reasoning. Such intuitions, however, might be unnecessarily limited. Cognitive science could try to obviate that problem by providing a unified theory of all reasoning that included moral reasoning as one variety. In that case, however, cognitive science would not have accounted for the specifically moral nature of moral reasoning. We would continue to need moral philosophy to provide such an understanding.

Ethics can also assist cognitive scientific studies of moral cognition by identifying the varieties and complexities of moral thinking. The variety of both morally normative and metaethical theories in Western moral philosophy is but a small sample of the wide-ranging moral variation around the world. Perhaps this moral pluralism reveals multiple possibilities for human moral cognition. Are all human moral conceptions necessarily exemplar based? Is all human moral reasoning necessarily a matter of pattern completion and pattern generalization?

Even if those developmental models are correct that claim that all human beings go through the same sequence of stages in the development of moral reasoning, still it appears that they do not all level off in adulthood with the same cognitive moral orientation. If adult moral cognition varied among cultures and among persons within the same culture, then what is cognitively impossible for some human beings might nevertheless be possible for others, and what is cognitively required for some might be unnecessary for others. If such were the case, then cognitive science would be unable to produce a uniform and generalizable account of the inner states and resources required for moral cognition. Widespread ethical plurality should signal to cognitive science (as to scientific inquiry in general) the importance of avoiding both simplistic conceptions of what morality is and oversimple theories to explain human moral cognition.

In addition to global moral pluralism, there is great diversity to moral reasoning within Western ethics alone. Moral philosophy articulates at least some of that variety. For example, moral philosophers know that the problem of abortion is not simply a matter of how women and girls should each deal with their own unwanted pregnancies; it is also a problem of what sorts of laws ought to govern a society of people who disagree about how women and girls should each deal with their unwanted pregnancies. Moral philosophers think that the sort of reasoning required to justify a moral norm or its implementation as social policy differs importantly from the sort required to apply it to particular cases. The exemplar-based account of moral reasoning will need to explain (or explain away) this distinction.

Moral philosophers also distinguish between high-level ethical principles, such as the categorical imperative, and lower-level moral rules, such as the rule against murder. They differentiate between, on one hand, religiously oriented moral frameworks that defer to the teachings of what are taken as moral authorities and, on the other hand, secular moral frameworks that promote moral problem solving through individual reflection guided by high-level principles. Moral philosophers differentiate between duties and consequences, between universalizing and attending to contextual particularity, indeed between moral thinking that focuses on exemplars, character, and virtue from moral thinking that focuses on rules and principles. If cognitive science is to provide a unified, all-encompassing account of the nature of moral reasoning, then it will have to deal with examples of each of these seemingly different sorts of moral cognition. Moral philosophers can usefully remind cognitive scientists of that vast variety.

Modern moral theory also prompts this question: If all moral cognition is exemplar based, why then would moral philosophers of the not-inconsiderable intellectual acumen of Kant and Mill, themselves presumably reasoning in an exemplar-based fashion, have so badly misconceived the nature of moral cognition—their own as well as that of others? Why would they have mistaken the nature of moral reasoning for a rule-based structure? (Why would so many philosophers have failed to notice this structure for cognition in general?) Is there something about the phenomenology or practice of moral reasoning that makes the rule-based account seem plausible?

It is reasonable to ask for some explanation of why rule-based theories have been so prevalent in modern Western ethics. The many philosophers who still think they gain moral understanding by thinking in a Kantian or utilitarian manner provide important counterexamples to which cognitive science should attend while exploring its exemplar-based model. Thus, rule-based ethical theories and their proponents provide an important perspective from which to appraise the plausibility of accounts of moral reasoning provided to ethics by cognitive science.

Furthermore, ethics and social theory reveal important methodological and conceptual controversies surrounding the nature of moral phenomena that are relevant to its scientific study. Scheman, for example, challenges the sort of psychological individualism that would treat emotions as nothing more than internal states of individuals. In its place, she defends a social constructionist account of emotions as socially meaningful patterns of thought, feeling, and behavior. Her defense also alerts us to the plurality of interpretations to which such moral phenomena as emotion are subject. The terms in which particular emotions (and all other mentalistic phenomena) are characterized can vary relative to standpoint and perspective. Describing certain emotions in a subject group as paranoia rather than anger may well express the observer's privileged and dominant cultural status. Until our moral vocabulary has been enriched by a dialogue among the members of what Scheman calls a "sufficiently democratic epistemic community," neither cognitive scientists nor moral philosophers, for that matter, can be sure that we are not mischaracterizing the cognitive moral phenomena that we are investigating. Ethics and social theory offer tools and resources that can help us to incorporate a recognition of moral diversity into our very theorizing about moral cognition.

Third, and finally, there are normative moral issues raised by the very practices of cognitive science that moral philosophy can helpfully address. In particular, the human and animal experimentation that goes on in cognitive science raises serious ethical questions. Should monkeys be subjected to mutilating and unanaesthetized surgery as part of research experiments? Should human beings ever be misinformed about the nature of experiments in which they are asked to participate? Ethics, especially biomedical ethics, has something to say about these sorts of issues in the ethics of research.

There is a common and regrettable tendency for scientific researchers, biomedical, cognitive, or otherwise, to avoid public discussion of these troubling ethical questions about their own professional practices. Often this disregard manifests the researcher's own overarching commitment to the scientific enterprise and its current norms of permissible research. Moral philosophers sometimes challenge the foundational moral legitimacy of current research practices. Scientists, in their haste to get on with their investigations, may wish to disregard those criticisms, may indeed regard them as bothersome interferences with the autonomy of science. This response, though understandable, is lamentable.

Science is a social enterprise supported by societal resources. Scientific experiments utilizing human or animal subjects pose all the same moral questions that are raised by any human activities that involve such morally salient features as the infliction of pain on sentient creatures. Scientists should not neglect these questions while forging ahead with experiments on live subjects. At the same time, trying to answer these questions without consulting the ideas of moral philosophers on the topic would be to rely on folk ethical resources that may be less articulated and nuanced on the subject of scientific research than the work of moral philosophers, who have recently devoted considerable attention to those issues. These ethical investigations provide yet another resource that ethics has to offer cognitive science.

Taken together, the contributions gathered in this book amply illustrate the power of cognitive scientific tools to illuminate important issues in ethics and moral philosophy. Conversely, work in ethics and moral philosophy can help cognitive scientists to see familiar problems (for example, that concerning the relation between language and cognition) in new and suggestive ways. And attention to the moral domain can help them gauge the generality and power of specific conceptions (such as the prototype-activation model) of human cognition.

A fully satisfying union between cognitive science and ethical theory will depend on a satisfactory resolution of at least two outstanding issues: the relationship between individual cognitive activity and the social dimensions of human knowledge and activity, and the relationship between the merely descriptive and the clearly normative. Such issues are at or near the surface of many of the discussions in this book. By attending to these and related concerns, the essays in this book break new ground in the emerging collaboration between ethics and cognitive science.

Note

1. Henceforth, names appearing in parentheses without dates refer to chapters in this book.

References

Churchland, P. M. 1989. *A Neurocomputational Perspective.* Cambridge, Mass.: MIT Press.

Clark, A. 1993. *Associative Engines: Connectionism, Concepts and Representational Change.* Cambridge, Mass.: MIT Press.

Clark, A. 1989. *Microcognition.* Cambridge, Mass.: MIT Press.

Clark, A., and Karmiloff-Smith, A. 1993. "The Cognizer's Innards: A Psychological and Philosophical Perspective on the Development of Thought." *Mind and Language* 8:487–519.

Flanagan, Owen. 1991. *Varieties of Moral Personality.* Cambridge, Mass.: Harvard University Press.

Fodor, J. 1987. *Psychosemantics.* Cambridge, Mass.: MIT Press.

Fodor, J. 1975. *The Language of Thought.* New York: Crowell.

Frankena, William. 1939. "The Naturalistic Fallacy." *Mind* 48:464–477. Reprinted in *Theories of Ethics,* edited by Philippa Foot, Oxford: Oxford University Press, 1967.

McClelland, J., Rumelhart, D., and the PDP Research Group. 1986. *Parallel Distributed Processing: Explorations in the Microstructure of Cognition.* Cambridge, Mass.: MIT Press.

Moore, G. E. 1903. *Principia Ethica*. Cambridge: Cambridge University Press.

Rosch, E. 1973. "On the Internal Structure of Perceptual and Semantic Categories." In T. Moore, ed., *Cognitive Development and the Acquisition of Language*. New York: Academic Press.

Searle, John. 1964. "How to Derive 'Ought' from 'Is.'" *Philosophical Review* 73:43–58. Reprinted in *Theories of Ethics*, edited by Philippa Foot, Oxford: Oxford University Press, 1967.

Sejnowski, T., and Rosenberg, C. 1987. "Parallel Networks That Learn to Pronounce English Text." *Complex Systems* 1:145–168.

Smith, E., and Medin, D. 1981. *Categories and Concepts*. Cambridge, Mass.: Harvard University Press.

Part I

Ethics Naturalized?

Chapter 2

Ethics Naturalized: Ethics as Human Ecology

Owen Flanagan

Ethics: Modern, Antimodern, and Postmodern

There are issues that separate ethical nonnaturalists from naturalists—issues such as whether ethics has theological foundations and whether a voluntarist conception of free will is, or is not, necessary for a proper conception of moral responsibility. There are also issues that divide naturalists from each other as well as from nonnaturalists: issues of the aim(s) of moral life, issues about the cognitive status of moral statements, about the existence of moral properties, about relativism, convergence, and moral progress.

In this chapter I sketch and defend a particular conception of ethics naturalized, one that explains the role of psychology, and the other human sciences in ethics, and explains why, despite there being ethical knowledge, it does not, as a matter of course, display either convergence or progress in the way some science does.

Modernism is the name usefully given to the view, inspired by enlightenment optimism, that modernity, committed as it is to rationality, is destined to yield convergence and progress in areas where human wisdom is to be had. In part this will be done by discovering the right general-purpose moral theory.

Bernard William's recent work contrasting Greek ethics with Kantian ethics is antimodern (as is his antimoral theory work).[1] The Greek concept of shame tied morality to where it should be tied—to being seen naked, as it were, in the eyes of others. The modern moral "ought" removes us from other persons to reason's sacred realm—a secular repeat of religious transcendentalism. Postmodernists are an unruly breed, and one can be antimodern—Williams and MacIntyre are in different ways—without being postmodern. But, if for no other reason than provocation, I claim that the conception I will be recommending is postmodern. It gives up both the idea that time, reason, and experience taken together reliably yield knowledge in moral and political life, and, at the same time, it abandons the concept of "reason" that is questioned, without being completely abandoned, by the antimodernists. It does so because contemporary mind science has no place for a faculty with the architectural features of reason, as traditionally construed. The architecture of mind, as it is emerging in contemporary mind science, is more like Frank Gehry's postmodern house in Santa Monica than it is like the modern British Museum or even the high-modern Seagram's Building in New York.[2] Why I say this will emerge in the course of the argument.

Why Ethics Naturalized Is Not Ethics Psychologized

In "Epistemology Naturalized," Quine suggested that epistemology be assimilated to psychology.[3] The trouble with this idea is apparent. Psychology is not in general concerned with norms of rational belief but with the description and explanation of mental performance and mentally mediated performance and capacities.

The right way to think of epistemology naturalized is not, therefore, one in which epistemology is a "chapter of psychology," but rather to think of naturalized epistemology as having two components: a descriptive-genealogical-nomological (d/g/n) component and a normative component.[4] Furthermore, not even the d/g/n component will consist of purely psychological generalizations, for much of the information about actual epistemic practices will come from biology, cognitive neuroscience, sociology, anthropology, and history—that is, the human sciences broadly construed. More obviously, normative epistemology will not be part of psychology, for it involves the gathering together of norms of inference, belief, and knowing that lead to success in ordinary reasoning and in science. And the evolved canons of inductive and deductive logic, statistics and probability theory, most certainly do not describe actual human reasoning practices. These canons—for example, principles governing representative sampling and warnings about affirming the consequent—come from abstracting successful epistemic practices from unsuccessful ones. The database is, as it were, provided by observation of humanity, but the human sciences do not (at least as standardly practiced) involve extraction of the norms. So epistemology naturalized is not epistemology psychologized.[5] But since successful practice—both mental and physical—is the standard by which norms are sorted and raised or lowered in epistemic status, pragmatism reigns.

The same sort of story holds for a naturalistic ethics. Naturalistic ethics will contain a d/g/n component that will specify certain basic capacities and propensities of *Homo sapiens*, such as sympathy, empathy, egoism, and so on, relevant to moral life. It will explain how people come to feel, think, and act about moral matters in the way(s) they do. It will explain how and in what ways moral learning, engagement, and response involve the emotions. It will explain what moral disagreement consists of and why it occurs, and it will explain why people sometimes resolve disagreement by recourse to agreements to tolerate each other without, however, approving of each other's beliefs, actions, practices, and institutions. It will tell us what people are doing when they make normative judgments. And finally, or as a consequence of all this, it will try to explain what goes on when people try to educate the young, improve the moral climate, propose moral theories, and so on.

Every great moral philosopher has put forward certain d/g/n claims in arguing for substantive normative proposals, and although most of these claims suffer from sampling problems and were proposed in a time when the human sciences did not exist to test them, they are almost all testable; indeed some have been tested.[6] For example, here are four claims familiar from the history of ethics that fit the bill of testable hypotheses relevant to normative ethics:

He who knows the good does it.
If you (really) have one virtue, you have the rest.
Morality breaks down in a roughly linear fashion with breakdowns in the strength and visibility of social constraints.
In a situation of profuse abundance, innate sympathy and benevolence will "increase tenfold," and the "cautious jealous virtue of justice will never be thought of."

Presumably, how the d/g/n claims fare matters to the normative theories, and would have mattered to their proposers. Nonetheless, no important moral philosopher, naturalist or nonnaturalist, has ever thought that merely gathering together all relevant descriptive truths would yield a full normative ethical theory. Morals are radically underdetermined by the merely descriptive, the observational, but so too, of course, are science and normative epistemology. All three are domains of inquiry where ampliative generalizations and underdetermined norms abound.

The distinctively normative ethical component extends the last aspect of the d/g/n agenda. It will explain why some norms—including those governing choosing norms—values, virtues, are good or better than others. One common rationale for favoring a norm or set of norms is that it is suited to modify, suppress, transform or amplify some characteristic or capacity belonging to our nature—either our animal nature or our nature as socially situated beings. The normative component may try to systematize at some abstract level the ways of feeling, living, and being that we, as moral creatures, should aspire to. But whether such systematizing is a good idea will itself be evaluated pragmatically.

Overall, the normative component involves the imaginative deployment of information from any source useful to criticism, self and/or social examination, formation of new or improved norms, and values: improvements in moral educational practices, training of moral sensibilities, and so on. These sources include psychology, cognitive science, all the human sciences, especially history and anthropology, as well as literature, the arts,[7] and ordinary conversation based on ordinary everyday observations about how individuals, groups, communities, nation-states, the community of persons, or sentient beings are faring.[8]

The standard view is that descriptive-genealogical ethics can be naturalized but that normative ethics cannot. One alleged obstacle is that nothing normative follows from any set of d/g/n generalizations. Another alleged obstacle is that naturalism typically leads to relativism, is deflationary, and/or is morally naive. It makes normativity a matter of power—either the power of benign but less than enlightened socialization forces, or the power of those in charge of the normative order, possibly fascists or Nazis or moral dunces.

Both the standard view and the complaint about naturalistic normative ethics turn on certain genuine difficulties with discovering how to live well and with a certain fantastical way of conceiving of what is "*really* right" or "*really* good." But these difficulties have everything to do with the complexities of moral life—of life, period—and have *no* bearing whatsoever on the truth of ethical naturalism as a metaphysical thesis *or* on our capacity to pursue genuinely normative and critical ethical inquiry.

A Minimalist Credo for the Ethical Naturalist

It will be useful to provide a minimalist credo for the ethical naturalist. First, ethical naturalism is nontranscendental in the following respect: it will not locate the rationale for moral claims in the a priori dictates of a faculty of pure reason—there is no such thing—or in divine design. Because it is nontranscendental, ethical naturalism will need to provide an "error theory" that explains the appeal of transcendental rationales and explains why they are less credible than pragmatic rationales, possibly because they are disguised forms of pragmatic rationales. Second, ethical naturalism will not be the slightest bit unnerved by open question arguments or by allegations of fallacious inferences from is to ought. With regard to open question problems, ethics naturalized need not be reductive, so there need be no attempt to define "the good" in some unitary way such that one can ask the allegedly devastating question: "But is that which is said to be 'good,' good?" To be sure, some of the great naturalists, most utilitarians, for example, did try to define the good in a unitary way. This turned out not to work well, in part because the goods at which we aim are plural and resist a unifying analysis. But second, the force of open question arguments fizzled with discoveries about failures of synonymy across the board—with discoveries about the lack of reductive definitions for most interesting terms.

With regard to the alleged is-ought problem, the smart naturalist makes no claims to establish demonstratively moral norms. He or she points to certain practices, values, virtues, and principles as reasonable based on inductive and abductive reasoning.

Third, ethical naturalism implies no position on the question of whether there really are, or are not, moral properties in the universe in the sense debated by moral realists, antirealists, and quasi-realists. The important thing is that moral claims can be rationally supported, not that all the constituents of such claims refer or fail to refer to "real" things. Fourth, ethical naturalism is compatible a priori, as it were, with cognitivism or noncognitivism. Naturalism is neutral between the view that "gut reactions" or, alternatively, deep and cautious reflection, are the best guide in moral life. Indeed, ethical naturalism can give different answers at the descriptive level and the normative level. Emotivism might be true as a descriptive psychosocial generalization, while a more cognitivist view might be favored normatively.[9]

Progress, Convergence, and Local Knowledge

So far I have argued that neither epistemology naturalized nor ethics naturalized will be a psychologized discipline *simpliciter*; neither will be a "chapter of psychology." Both will make use of information from all the human sciences and in the case of ethics from the arts as well, for the arts are a way we have of expressing insights about our nature and about matters of value and worth. The arts are also—indeed, at the same time and for the same reasons—ways of knowing, forms of knowledge. Norms will be generated by examining this information in the light of standards we have evolved about what guides or constitutes successful practice.

I have also sketched a brief credo for the ethical naturalist—one that captures what at a minimum anyone needs to be committed to to count as one, a credo that leaves open the possibility of different answers to question about moral realism and antirealism and about cognitivism and noncognitivism.[10]

So far I have been working with the proposal that ethics naturalized can go the same route as epistemology naturalized. But things that are alike in many respects are not alike in all respects. Quine himself is skeptical of a robust, naturalized ethics, arguing instead that ethics is "methodologically infirm" compared to science[11] and that both naturalized epistemology and science, unlike ethics, have an unproblematic, uncontroversial end, truth and prediction—a goal that is simply "descriptive."[12]

These ways of characterizing the situation of ethics on the one side and epistemology and science on the other are both off target. Doing science at all and then doing it the ways we do it, and for the reasons we do it, involve a host of normative commitments. Nonetheless, I do think we need an account for the often observed fact that ethical wisdom, insofar as there is such a thing, fails to reach convergence in the way epistemic norms and scientific knowledge do. By "science" I mean, for ease of argument, to restrict myself to areas like anatomy, physiology, inorganic and organic chemistry, and physics without general relativity or quantum physics. Convergence in epistemology and science is, over long periods of time, thought to be correlated with, tracking ever more closely, "the right ways to collect and utilize evidence" and "the truth about the way things are." So there is both convergence and progress.

In the rest of the chapter, I explain why the lack of convergence about ethical matters, insofar as there is such lack, and the difficulty with applying the concept of progress to moral knowledge is something we ought to accept and that it harms neither the naturalistic project nor the idea that there is, or can be, ethical knowledge. In effect, I will argue for a certain asymmetry between epistemology naturalized and science, on the one side, and ethics, on the other, although I have been emphasizing some of their similarities.

A way to state my general idea is this: the links between any inquiry and the convergence and/or progress such inquiry yields is determined in large part by the degree of contingency and context dependency the target domain exhibits and the way the end or ends of inquiry are framed. The basic sciences, due to the univocality and consistency of their ends, and to the nonlocal nature of the wisdom they typically seek, converge more and give evidence of being progressive in ways that ethics does not. The explanation has to do with the fact that the ends of ethics are multiple, often in tension with each other, and the wisdom we seek in ethics is often local knowledge—both geographically local and temporally local, and thus convergence is ruled out from the start. Indeed, even in one and the same geographic and temporal location, different groups occupying that location might hold different norms and values, and this lack of local convergence might be sensibly thought to be good, not simply tolerable.[13]

Some naturalists take the refinement and growth of moral knowledge over time—for the individual and the culture—too much for granted. For example, most neo-Piagetian models emphasize the invisible hand of some sort of sociomoral assimilation-accommodation mechanism in accounting for the refinement of moral knowledge over time. Others emphasize the visible processes of aggregation of precedent, constraint by precedents, and collective discussion and argument that mark sociomoral life. One can believe that moral knowledge exists without thinking, that it grows naturally over time, that it has increased in quality and quantity in the course of human history.

Moral Network Theory

The best way to explain these points is by discussing a specific account of moral learning recently proposed by a fellow pragmatist and naturalist, Paul Churchland.[14] I call Churchland's view "moral network theory."[15] First, I will offer a brief reconstruction of the d/g/n component of moral network theory. Then I will address the tricky problem of extrapolating from moral network theory as a credible model of moral learning, habit and perception formation, moral imagination, and reasoning to the conclusion that the process yields convergence, leads to progress, and yields objective knowledge, now pragmatically understood. Still, I will insist that the worries I raise offer no solace to those who think that the naturalist has no resources, no perspective, from which to make judgments about what is good and right.

Acquiring knowledge, according to moral network theory, "is primarily a process of learning how: how to recognize a wide variety of complex situations and how to respond to them appropriately. Through exposure to situations—situations that include talk—moral perception, cognition, and response develop and are refined. According to moral network theory, there is a straightforward analogy between the way a submarine sonar device that needs to learn to distinguish rocks from mines might acquire the competence to do so and the way a human might acquire moral sensitivities and sensibilities.

One way to teach the mine-rock device would be simply to state the rule specifying the necessary and sufficient characteristics of rocks and mines. The trouble is that these are not known (indeed it is part of the mine producers' job to make them as physically nondistinct as possible. Despite these efforts at disguise, there are bound to be (or so we hope) subtle features that distinguish mines from rocks, so it would be good if the device could be trained in a situation where it starts by guessing mine or rock and then, by being clued into the accuracy of its guesses, develops a profile for recognizing rocks from mines. Indeed this can be done with connectionist networks (figure 2.1). Eventually the mine-rock detector (which, of course, never becomes perfect at its job) comes to be able to make judgments of kind very quickly, based on a small number of features, and it responds accordingly. One might compare the acquisition of the skill involved in that of learning to recognize a partially concealed toy or animal for the kind of toy or animal it is (figure 2.2).

According to moral-network theory, the fundamental process is the same in the case of moral learning. Children learn to recognize certain

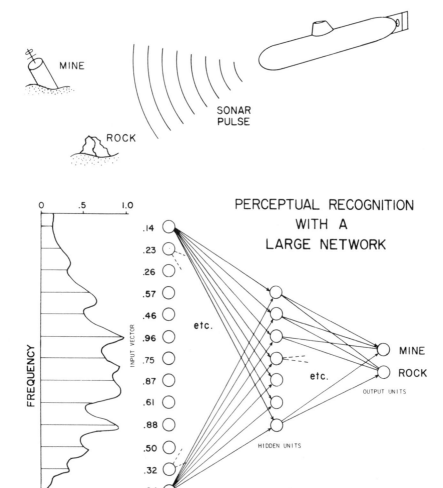

Figure 2.1
Mine-rock detector and network. Source: P. M. Churchland, *Matter and Consciousness*, rev. ed. (Cambridge, Mass.: MIT Press, 1988).

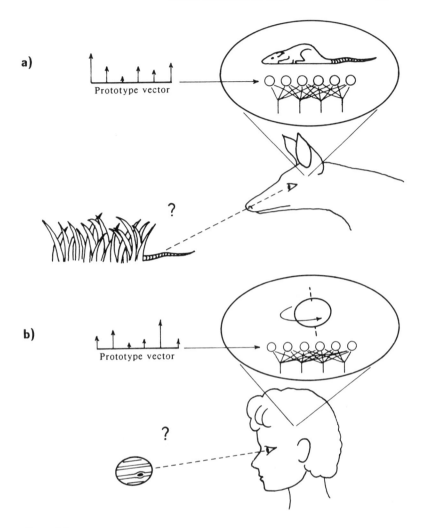

Figure 2.2
Ampliative perceptual recognition. Source: P. M. Churchland, *A Neurocomputational Perspective: The Nature of Mind and the Structure of Science* (Cambridge, Mass.: MIT Press, 1988).

prototypical kinds of social situations, and they learn to produce or avoid the behaviors prototypically required or prohibited in each." Children come to *see* certain distributions of goodies as a *fair* or *unfair distribution*. They learn to recognize that a found object may be someone's property, and that access is limited as a result. They learn to discriminate *unprovoked cruelty*, and to demand or expect punishment for the transgressor and comfort for the victim. They learn to recognize a *breach of promise*, and to howl in protest. They learn to recognize these and a hundred other prototypical social/moral situations, and the ways in which the embedding society generally reacts to those situations and expects them to react."[16]

The observations here are no less ampliative (although they are possibly more complex) than judging that *it looks like rain*, that the *peach is not ripe yet*, that *the air conditioning doesn't sound right*, that *Jane is shy*, that *the fans are in a frenzy*, that *Bob is impatient*. These are the sorts of discriminative judgments we learn to make with ease, but not one of them is an observation sentence.

To indicate how complex the learning problem the child faces in just one domain, consider learning to tell the truth—learning, as we say, that honesty is the best policy. One thing we know for sure is that we do not teach, nor do we want to teach, the child that he has a categorical obligation to "tell the truth whenever the truth can be told."

With respect to truth telling the child needs to learn:

1. What truth telling is (why joking and fairy tales are not lies).
2. What situations call for truth telling.
3. How to tell the truth.

Perhaps this sounds simple; it is a short list, after all. But consider the fact that we consider it important that novices learn to distinguish among situations that seem to call for truth telling, not all of which do—or which do but not in the same way. Consider just four kinds of situations and what they require in terms of discrimination and response:

1. *Situations that call for straightforward truth telling:* "The cookies were for dessert. Did you eat them all, Ben?"
2. *Situations that call for tact:* "So, Ben, you are enjoying school, aren't you?" (said by teacher to child in front of parents).
3. *Situations that call for kind falsehoods/white lies:* "Kate, I got my hair cut a new way for the party tonight. How do you like it?" (one preteen to another). "Kate, don't you think I'm getting better at soccer?" (said by one teammate to another—and supposing Kate does not think Emily has improved one bit over the season).
4. *Situations that call for lying/misinformation, depending on who is asking:* "Little boy, what is your address?" (asked by a stranger).

How exactly a child or an adult responds to a novel moral situation "will be a function of which of her many prototypes that situation activates, and this will be a matter of the relative similarity of the new situation to the various prototypes on which she was trained." Situations will occasionally be ambiguous, and there will be disagreements about what is occurring. "What seems a case of unprovoked cruelty to one child can seem a case of just retribution to another."[17] Moral ambiguity creates mental cramps of various sorts that lead to reflection, discussion, and argument, as does disagreement among persons about how to describe a certain act or situation. These, in turn, lead to prototype adjustment.

Two comments are in order before we proceed further. First, it should be emphasized, in case it is not obvious, that the moral network theorist is using the concept of a prototype in a very broad but principled sense. Indeed, Churchland argues that the prototype activation model is general enough to account for perceptual recognition and scientific explanation—for example, inference to the best explanation is (roughly) activation of "the most appropriate prototype vector" given the evidence.[18] So prototypes cover both knowing how and knowing that. And what Churchland calls "social-interaction prototypes underwrite ethical, legal, and social-etiquette explanations."[19]

We explain, for example, why the Czech driver stops to help the man with the flat tire by explaining that there are Good Samaritan laws—rules to be applied in certain situations—in the Czech republic. Conversely, the existence of the relevant social expectations expressed in such laws explains why the Czech driver sees a person with a flat as someone in need of his help and sees helping as something he ought to do—perhaps for prudential reasons, perhaps because samaritanism is deeply ingrained. The sort of sociomoral learning that takes place in the Czech republic, but not in (most parts of) America, means that Czech drivers just stop to help people stuck on the side of the road and that seeing drivers stuck on the side of the road engages certain feelings—either fear of the law if one is observed not stopping or a moral pull to help if the ingrained disposition involves some sort of genuine compassion or fellow feeling.

The second comment is that the prototype activation model designed to explain moral learning must explain our notion of morality itself. Morality does not pick out a natural kind. What is considered moral, conventional, personal, and so on depends on the complex set of practices, including practices involving the use of certain terms, that have evolved over time in different communities. If one asks for a definition of "moral" in our culture, it will be framed initially in terms of rules designed to keep humans from harming other humans—in terms of conflict avoidance and resolution. If, however, one looks to the meanings that reveal themselves in ascriptive practices and in response to verbal probes of a more complex

sort, one will also find that we think of the moral as having an intrapersonal component. That is, there are norms governing personal goodness, improvement, and perfection, as well as norms governing self-respect
and self-esteem, that are not reducible to conformity to interpersonal moral norms.[20] This point will matter for what I say later.[21]

So far I have sketched moral network theory as a descriptive-
genealogical-nomological theory. It is a credible theory, and possibly true.
Indeed, there are two aspects of the view that I find particularly compelling and fecund. The first is the idea that moral responsiveness does not
(normally) involve deployment of a set of special-purpose rules or algo-
rithms that are individually applied to all, and only, the problems for which
they are designed specifically. Nor does moral responsiveness normally
involve deployment of a single general-purpose rule or algorithm, such as
the principle of utility or the categorical imperative, designed to deal with
each and every moral problem. Moral issues are heterogeneous in kind,
and the moral community wisely trains us to possess a vast array of moral
competencies suited—often in complex combinations and configurations
—to multifarious domains, competencies that in fact and in theory resist
unification under either a set of special-purpose rules or under a single
general-purpose rule or principle. This is what I meant earlier when I
suggested that contemporary models of mind do not leave any obvious
place for a traditional faculty of reason or, for that matter, for a process of
rational rule governance as traditionally conceived. The architecture of
cognition is neither that of the modern faculty psychologist nor of the
high modernist, rule-following, serial von Neumann device. Connectionist
"reasoning," neural network "reasoning," is at best a *façon de parler*, since
really all such systems can do is rooted in recognizing and responding to
patterns. Networks do pattern recognition, as it were; they do not "reason" in any traditional sense of the term. The architecture of mind is
postmodern.

The second feature of the view that I like is this: the theory is, as I call
it, a moral network theory, but the total network comprises more than the
neural nets that contain the moral knowledge a particular individual possesses. Whatever neural net instantiates (or is disposed to express) some
segment of moral knowledge, it does so only because it is "trained" by a
community. The community itself is a network providing constant feedback to the human agent.

The neural network that underpins moral perception, thought, and action is created, maintained, and modified in relation to a particular natural
and social environment. The moral network includes but is not exhausted
by the dispositional states laid down in the neural nets of particular
individuals.

Normativity

It is time to see how moral network theory—as an exemplar of the sort of d/g/n theory likely to be offered up by the naturalist—can handle the normative issue. Churchland himself raises the worry in vivid terms.

> What is problematic is whether this process amounts to the learning of genuine Moral Truth, or to mere socialization. We can hardly collapse the distinction, lest we make moral criticism of diverse forms of social organization impossible. We want to defend this possibility, since ... the socialization described above can occasionally amount to a cowardly acquiescence in an arbitrary and stultifying form of life. Can we specify under what circumstances it will amount to something more than this?[22]

Churchland claims that "an exactly parallel problem arises with regard to the learning of Scientific Truth" since almost no scientific learning is first-hand, and even that which is occurs within a research tradition with a particular orientation, set of methods, and so on. Nonetheless, despite overblown views of Scientific Truth and the Correspondence to the Real held by certain scientistic thinkers and hard-core metaphysical realists, Churchland insists that "there remains every reason to think that the normal learning process, as instanced both in individuals and in the collective enterprise of institutional science, involves a reliable and dramatic increase in the amount and the quality of the information we have about the world."[23] And, he claims, the same is true of moral knowledge:

> When such powerful learning networks as humans are confronted with the problem of how best to perceive the social world, and how best to conduct one's affairs within it, we have equally good reason to expect that the learning process will show an integrity comparable to that shown on other learning tasks, and will produce cognitive achievements as robust as those produced anywhere else. This expectation will be especially apt if, as in the case of "scientific" knowledge, the learning process is collective and the results are transmitted from generation to generation. In that case we have a continuing society under constant pressure to refine its categories of social and moral perception, and to modify its typical responses and expectations. Successful societies do this on a systematic basis.[24]

Churchland then asks:

> Just what are the members of the society learning? They are learning how best to organize and administer their collective and individual affairs. What factors provoke change and improvement in their typical

categories of moral perception and their typical forms of behavioral response? That is, what factors drive moral learning? They are many and various, but in general they arise from the continuing social experience of conducting a life under the existing moral framework. That is, moral learning is driven by social experience, often a long and painful social experience, just as theoretical science is driven by experiment. Moral knowledge thus has just as genuine a claim to objectivity as any other kind of empirical knowledge. What are the principles by which rational people adjust their moral conceptions in the face of unwelcome social experience? They are likely to be exactly the same "principles" that drive conceptual readjustment in science or anywhere else, and they are likely to be revealed as we come to understand how empirical brains actually do learn.[25]

The core argument is this: moral knowledge is the result of complex socialization processes. What keeps moral socialization from being "mere" socialization has to do with several features of sociomoral (and scientific) life. "Mere" socialization is socialization toward which no critical attitude is taken, for which there are no rational mechanisms that drive adjustment, modification, and refinement. The reason moral socialization is not (or need not be) "mere" socialization has to do with the fact that there are constraints that govern the assessment and adjustment of moral learning. We are trying to learn "how best to organize and administer [our] collective and individual affairs." Social experience provides feedback about how we are doing, and reliable cognitive mechanisms come into play in evaluating and assessing this feedback. So there are aims, activities to achieve these aims, feedback about success in achieving the aims, and reliable cognitive mechanisms designed to assess the meaning of the feedback and to make modifications to the activities.

These reliable cognitive mechanisms include individual reflection and collective reflection and conversation. Now it is important to understand what our reflective practices are ontologically. They do not involve the deployment of some rarefied culture-free faculty of reason. Dewey puts it best: "The reflective disposition is not self-made nor a gift of the gods. It arises in some exceptional circumstances out of social customs ... [w]hen it has been generated it establishes a new custom."[26] We need to recognize that our practices of reflection are themselves natural developments that emerged in large measure because they were useful.[27] Our reflective practices, from individual reflection to collective conversation and debate, are themselves customs that permit "experimental initiative and creative invention in remaking custom."[28] The choice "is not between a moral authority outside custom and one within it. It is between adopting more or less intelligent and significant customs."[29]

I think of these Deweyan points as extensions of moral network theory's general line about moral knowledge.[30] But this idea that to be rational,[31] self-conscious, and critical are developments of natural capacities, they are things we learn to be and to do, not transcendental capacities we simply have, is still not well understood or widely accepted. But it is true. Critical rationality is a perfectly natural capacity displayed by *Homo sapiens* socialized in certain ways—fancy, yes, but nothing mysterious metaphysically.

So far I have tried to extend the picture of moral learning favored by moral network theory, indicating how it can account for complex prototype activation and for our rational practices themselves as socially acquired and communally circumscribed in structure and content.

Moral Progress and Moral Convergence

I now speak about two concerns with the picture of normativity that Churchland paints in the name of moral network theory. The first concern has to do with the strong similarity alleged to obtain between scientific and ethical knowledge, the second with the problematic view of moral objectivity, convergence, and progress that the too-tidy analogy supports. My main reservation has to do with Churchland's overly optimistic attitude about the capacities of the moral community to arrive at high-quality moral knowledge[32] and with the failure to emphasize sufficiently the local nature of much moral knowledge.

First, there is the overly optimistic attitude about moral progress. Both Paul Churchland and I are pragmatists, so I agree with him completely when he says that "the quality of one's knowledge is measured not by any uniform correspondence between internal sentences and external facts, but by the quality of one's continuing performance."[33] I also agree with him when he writes of sociomoral knowledge acquisition that "what the child is learning in this process is the *structure of social space* and *how best to navigate one's way through it.* What the child is learning is practical wisdom: the wise administration of her practical affairs in a complex social environment. This is as genuine a case of learning about objective reality as one finds anywhere. It is also of fundamental importance for the character and quality of any individual's life, and not everyone succeeds equally in mastering the relevant intricacies."[34]

Here is the rub: what the child learns about "the wise administration of her practical affairs" is a complex mix of moral knowledge, social savvy, and prudential wisdom. These things overlap in important ways, but they do not overlap in all respects, and they are worth distinguishing. Once they are distinguished, it is easy to see that these things are typically in some tension with each other. The demands of social success and prudence can compete with each other and with the demands of morality. If most knowing is knowing "how," then what the sociomoral community must do is teach people how to resolve such conflicts. Communities of possessive individualists learn one form of resolution, Mother Teresa's nuns a different one. Practical success may come to both communities as they see it, but one might argue that moral considerations are hardly improved or developed equally in both communities. This may be fine, the way things should go, all things considered. But it is important not to think that when Churchland writes that "we have a continuing society under constant pressure to refine its categories of social and moral perception, and to modify its typical responses and expectations" that the *"constant pressure to refine"* is primarily working on the moral climate. The reason we should not think this is simply because there are too many competing interests besides moral ones vying for control of our sociomoral responses: simple self-interest, prudence, concern with economic, social, and sexual success, and much else besides. Furthermore, it is in the interest of different individuals and social groups to train others to believe (and even to make it true up to a point) that their being well and doing well involves conforming to norms that produce disproportionate good for the trainers.

A different but related point comes from reflecting on the ends of ethics. I certainly do not think that the ends of science are simple or unambiguous, but I suspect that they do not display the inherent tension that the ends of ethics do. What I have in mind goes back to my comments about the two sides of ethics: the interpersonal, concerned with social stability, coordination, prevention of harms, and so on; and the intrapersonal, concerned with individual flourishing, personal goodness, and the like. The tension between impartial moral demands and what conduces to individual flourishing is ubiquitous. Moral ambiguity is endemic to the first steps of ordinary moral life. We wonder how spending so much on luxuries could possibly be justified given the absolute poverty that exists in many places in the world. Much recent philosophical attention has been given to this tension between the more impartial demands of morality as we are exposed to it and the goods of personal freedom, choice, and integrity that we also learn.[35] There are conceivable ways to remove the tension, remove just one of the dual ends of moral life from the way we construct our conception of it. But this seems an unappealing idea. It seems best to leave the tension as is. It is real, and more good comes from having to confront it again and again than could come from simply stipulating it away. But one consequence of leaving the causes of moral ambiguity in place is that we will often rightly feel a certain lack of confidence that we have done the right thing. This suggests that we are often, at least in the moral sphere, in situations where we have no firm knowledge to go on.

A related consequence of the fact that the very aims of ethics are in tension is this: interpersonal ethics requires that we cast the normative net widely and seek agreement about mechanisms of coordination and about what constitutes harm. The requirements of intrapersonal ethics create pressures to cast the net narrowly. We think it is good that different people find their good in different ways. It is good that the Amish construct Amish virtue the way they do, and it is good that some communities think that benevolence is the most important individual virtue, while others think that humility or peace of mind is. One problem that often occurs, and that this picture helps us understand, is that wide-net policies, such as U.S. abortion laws, can not only conflict with the ways some individuals and groups find ethical meaning but also undermine their sense of their own integrity by virtue of their complex commitment to both the larger culture and the values of their group.

The interpersonal side in casting its net more widely may well yield relatively global moral knowledge: it just is wrong to kill someone except in self-defense.[36] But the idea that different persons find their good in different ways will require acknowledgment that much, possibly most, ethical knowledge is local knowledge. It is knowledge possessed by a particular group—Catholics, the Amish, secular humanists, Hindus, Muslims—and it is hugely relevant to the quality of the lives of the members of the group that they live virtuously as they see it.[37]

Moral network theory, with its postmodern mental architecture, goes badly with naive enlightenment optimism about moral progress and convergence. Progress and convergence might come our way, but they will not come by way of some natural dialectic of reason (conceived again in the enlightenment sense) and experience. The network will no doubt yield some workable ways to "organize and administer" our "collective and individual affairs." But for reasons I have suggested, there is no remotely reliable guarantee that these "workable ways" will be morally workable. The pragmatist needs to be sensitive to different and conflicting senses of "what works."

In closing, I comment on Dewey's insight that "moral science is not something with a separate province. It is physical, biological, and historic knowledge placed in a humane context where it will illuminate and guide the activities of men."[38] What is relevant to ethical reflection is everything we know, everything we can bring to ethical conversation that merits attention: data from the human sciences, history, literature and the other arts, from playing with possible worlds in imagination, and from everyday commentary on everday events. Critique is perspectival, not transcultural or neutral; it originates in reflection from the point of view of this historical place and time.[39]

One lesson such reflection teaches is that if ethics is like any science or is part of any science, it is part of human ecology concerned with saying what contributes to the well-being of humans, human groups, and human individuals in particular natural and social environments.[40]

Critique, on the view I am defending, can be as radical as one wishes.[41] It is just that it will be perspectival. It will originate from here and now, and not from some neutral, transcultural, or transcendental perspective. When we engage in the projects of self-understanding and self-criticism by looking to other moral sources—within our tradition or from another culture—we obviously challenge our own modes of self-understanding. Less obviously, but no less truly, when we criticize an alternative way of being or acting, our own ways of being and acting are changed in the dialectic of critique. But all critique is immanent. *Its meaning and outcome is source dependent.*[42]

Thinking of normative ethical knowledge as something to be gleaned from thinking about human good relative to particular ecological niches will make it easier for us to see that there are forces of many kinds, operating at many levels, as humans seek their good; that individual human good can compete with the good of human groups and of nonhuman systems; and, finally, that only some ethical knowledge is global, most is local, and appropriately so. It might also make it seem less compelling to find ethical agreement where none is needed. Of course, saying what I have said is tantamount to affirming some form of relativism. I intend and welcome this consequence. But defending ethics naturalized, a pragmatic ethic conceived as part of human ecology seems like enough for now. So I will save the formulation and defense of the relativism I favor until some later time.[43]

Acknowledgments

Earlier versions of this chapter were given at the American Philosophical Association meetings in Los Angeles, and at the Mind and Morals Conference at Washington University in St. Louis, in March and April 1994, and in June at Monash University, Melbourne, Australia. I am grateful to many individuals for helpful comments and criticisms.

Notes

1. See B. A. O. William, *Shame and Necessity* (Berkeley: University of California Press, 1993). The antitheory side comes out most clearly in *Ethics and the Limits of Philosophy* (Cambridge: Harvard University Press, 1985). I think that the search for a general-purpose moral theory is a waste of time. The French postmodernists Foucault, Deleuze, and Lyotard reject moral theory as well. But they make the misstep from the premise that there is no single moral theory to the conclusion that moral discourse is to be shied away from.

2. F. Jameson, *Postmodernism* (Durham, N.C.: Duke University Press, 1991). Some identify postmodern architecture with crass commercialism. The Trump Tower is then the perfect icon. I am using postmodern here to get at the ideas that there are none of the usual rooms (certainly no room for pure reason) and that traditional ideas of function are abandoned or reconsidered. This is what is happening as we see another way to conceive of mental function than the serial, programmable, reasoning model that dominated cognitive science for its first twenty years. Its replacement, connectionist models that work but are not yet fully understood, seems to me, at any rate, like the Gehry house.

3. W. V. Quine, "Epistemology Naturalized," in *Ontological Relativity and Other Essays* (New York: Columbia University Press, 1969), pp. 69–90.

4. I am not sure whether there are laws of the human mind that are relevant to ethics, but the "nomological" holds a place for them should some turn up.

5. To my mind the best work in naturalized epistemology is that of Alvin Goldman. See *Epistemology and Cognition* (Cambridge: Harvard University Press, 1986), and *Liaisons: Philosophy Meets the Cognitive and Social Sciences* (Cambridge: MIT Press, 1992). Goldman never tries to derive normative conclusions from descriptive premises. Furthermore, he continually emphasizes the historical and social dimensions of epistemology in a way that Quine does not.

6. See Owen Flanagan, *Varieties of Moral Personality: Ethics and Psychological Realism* (Cambridge: Harvard University Press, 1991).

7. Richard Rorty convincingly suggests that the formulation of general moral principle has been less useful to the development of liberal institutions than has the gradual "expansion of the imagination" [through works] like those of Engels, Harriet Taylor and J. S. Mill, Harriet Beecher Stowe, Malinowski, Martin Luther King Jr., Alexis de Tocqueville, and Catherine MacKinnon. (Rorty, "On Ethnocentrism," in *Philosophical Papers* (Cambridge: Cambridge University Press, 1991), 1: 207.

8. Critics of naturalized ethics are quick to point out that notions like "flourishing," "how people are faring," "what works for individuals, groups, nation-states, the world, etc." are vague and virtually impossible to fix in a noncontroversial way. This is true. The pragmatist is committed to the requirement that normative judgments get filled out in conversation and debate. Criteria of flourishing, what works, and so on will be as open to criticism as the initial judgments themselves. It is hard, therefore, to see how the criticism *is* a criticism. The naturalist is open to conversational vindication of normative claims, she admits that her background criteria, cashed out, are open to criticism and reformulation, and she admits that phrases like "what works" and "what conduces to flourishing" are superordinate terms. Specificity is gained in more fine-grained discussion of particular issues. But in any cases, there is no ethical theory ever known, naturalist or nonnaturalist, that has not depended on abstract concepts. Thin concepts sometimes yield to thick concepts: "That's bad." "Why?" "Because it is immodest." Now one can and often does stop here. But one can go on in any number of directions: "Why is it immodest?" "Why should I care about immodesty?" And so on.

9. Bernard Williams thinks, as I do, that the descriptive underdetermines the normative. But there is a relevance relation. See his *Shame and Necessity* for an evaluation of the positive and negatives associated with shame versus guilt cultures. This is work that has—indeed that requires as its data base—sophisticated psychosocial-historical knowledge. Allan Gibbard's recent work—*Wise Choices, Apt Feelings: A Theory of Normative Judgment* (Cambridge: Harvard University Press, 1990)—links the descriptive and normative. Gibbard describes his "expressivist view" as noncognitivist, but I find it more useful to think of it, *qua* normative proposal, as cognitivist, as a rational reconstruction of what is involved in moral competence, where what is involved is expressing our allegiance to norms that have been vindicated by assessments in terms of aptness, in terms of what it makes sense to think, feel, believe, and so on.

10. In both the realism-antirealism cases and the cognitivist-noncognitivist case, different answers might be given at the descriptive and normative levels. J. L. Mackie, *Ethics: Inventing Right and Wrong* (New York: Penguin, 1977), is an example of a philosopher who thought that ordinary people were committed to a form of realism about values but were wrong. In spite of this, Mackie saw no problem with advocating utilitarianism as the best moral theory and in that sense was a cognitivist—a cognitivist antirealist, as it were.

11. W. V. O. Quine, "On the Nature of Moral Values," *Critical Inquiry* 5 (1979):471–480.

12. In a later paper—W. V. O. Quine, "Reply to Morton White," in *Philosophy of W. V. Quine*, ed. L. E. Hahn and P. A. Schilpp (La Salle, Ill.: Open Court, 1986)—than the one in which the methodological infirmity charge ("On the Nature of Moral Values") is foisted, Quine puts the way naturalized epistemology will work better that he does in "Epistemology Naturalized," where the assimilated-to-psychology proposal is made. He writes: "A word now about the status, for me, of epistemic values. Naturalization of epistemology does not jettison the normative and settle for the indiscriminate description of ongoing procedures. For me normative epistemology is a branch of engineering. It is the technology of truth-seeking, or, in a more cautiously epistemological term, prediction. Like any technology, it makes free use of whatever scientific findings may suit its purpose. It draws upon mathematics in computing standard deviation and probable error and in scouting the gambler's fallacy. It draws upon experimental psychology in exposing perceptual illusions, and upon cognitive psychology in scouting wishful thinking. It draws upon neurology and physics, in a general way, in discounting testimony from occult or parapsychological sources. There is no question here of ultimate value, as in morals; it is a matter of efficacy for an ulterior end, truth or prediction. The normative here, as elsewhere in engineering, becomes descriptive when the terminal parameter is expressed. We could say the same of morality if we could view it as aimed at reward in heaven."

I do not see how the argument can be made to work. To be sure, any socially enriched theory of what happiness or welfare or flourishing consists in will be irreducible to observation or, what is different, to basic facts about human nature. But it is hard to see how this constitutes a significant objection since, according to Quine, the slide from unreduced to unjustified is illegitimate. "There are electrons" is, after all, irreducible but not thereby unjustified. Furthermore, it is not as if our noninstrumental, socially enriched, moral conceptions are picked from thin air. In the first place, they invariably have some, possibly very remote, links to certain basic appetites and aversions. Second, even if we cannot reduce our ultimate values to something else, we can often say nontrivial things about them that have justificatory bearing. For example, we can sometimes say things about the ways in which some newly discovered good is good. This need not involve instrumental reduction as much as feature specification of the nature of the particular good. Third, even if we find that there are certain things, pain for example, about which we have little more to say than that they are just plain bad, this does not engender lack of justification—especially since science's bottoming out in similarly bald facts is taken by Quine to ground it. It is simply stipulative to suggest that the values that guide science and normative epistemology, as we know them, are unproblematic "ulterior ends"—some sort of "descriptive terminal parameter" specified by nature in the way our values would be specified by God if there was a God. We do science and epistemology in ways that show every sign of being driven by socially specific values. Furthermore, insofar as we can ground some sort of minimal transcultural "need to know," it will be grounded in certain basic features of persons as epistemically capable biological organisms, as well as in certain basic desires to live well—characteristics that may also drive our ethical constructions across social worlds.

The attempt to draw a distinction between the unproblematic ulterior ends of truth and prediction that guide science and epistemology and our ultimate ethical values that sit "unreduced and so unjustified" tries to mark a distinction where there is none and represents yet another remnant of positivistic dogma.

13. And even if there are shared values in this locale, moral knowledge might be domain specific. This is an idea that has been pressed in different ways by Virginia Held, *Rights and Goods* (Minneapolis: University of Minnesota Press, 1984); Michael Walzer, *Spheres of Justice* (New York: Basic Books, 1983); and Flanagan, *Varieties.*

14. Paul M. Churchland, *A Neurocomputational Perspective: The Nature of Mind and the Structure of Science* (Cambridge: MIT Press, 1989). See also Mark Johnson, *Moral Imagination: Implications of Cognitive Science for Ethics* (Chicago: Chicago University Press, 1993).

15. Owen Flanagan, "The Moral Network," in *The Churchlands and Their Critics*, ed. R. McCauley (London: Basil Blackwell, in press).

16. Churchland, *Neurocomputational Perspective*, p. 299.

17. Ibid., p. 300.

18. Ibid., p. 218.

19. Ibid., p. 216.

20. Flanagan, *Varieties.*

21. One issue I have not emphasized here, but discuss as length in ibid., is the way personality develops. One might think that it is a problem for a network theory that it will explain the acquisition of moral prototypes in such a way that character is just a hodgepodge of prototypes. I do not think this is a consequence of moral network theory. One idea is to emphasize the fact that the human organism is having its natural sense of its own continuity and connectedness reinforced by the social community. A self, a person is being built. The community counts on the individual's developing a sense of his own integrity, of seeing himself as an originator of agency, capable of carrying through on intentions and plans, and of being held accountable for his actions. The community, in effect, encourages a certain complex picture of what it means to be a person, and it typically ties satisfaction of moral norms to this picture. Identity involves norms, and the self is oriented toward satisfying these norms. This helps explain the powerful connections that shame, guilt, self-esteem, and self-respect have to morality. Morality is acquired by a person with a continuing identity who cares about norm satisfaction and about living well and good.

22. Churchland, *Neurocomputational Perspective*, p. 300.

23. Ibid., p. 301.

24. Ibid., pp. 301–302. Churchland continues the quoted passage this way: "A body of legislation accumulates, with successive additions, deletions, and modifications. A body of case law accumulates, and the technique of finding and citing relevant precedents (which are, of course, *prototypes*) becomes a central feature of adjudicating legal disputes." The law is a good example of a well-controlled normative domain where previous decisions produce weighty constraints and norms accumulate. But it is not obvious that the law and morality, despite both being normative, are alike in these respects. In the law, publicity, precedent, and stability are highly valued; indeed they are partly constitutive of the law. Public, stable norms, with precedents, are thought to be good, and essential to successful and reliable negotiation of the parts of life covered by the law. But it is also recognized that the goods of stability, publicity, and so on are perfectly compatible with irrational or immoral laws. So one point is that the aims of the law are not the same as those of ethics (ignorance of the law is no defense; ethical ignorance can be). Another issue arises in the first part of the quoted passage where Churchland frames the issue as one of humans' confronting "the problem of how best to perceive

the social world, and how best to conduct one's affairs within it." This is, I think, just the right way to see the problem. But there is a telling difference between, for example, learning about legal reality or how best to perceive the causal, temporal, spatial structure of the world and how to conduct one's affairs in this (these) world(s). The difference is simply this: these latter structures are relatively more stable and relatively more global than the moral structure of the world is. This is one reason that Piagetian developmental stage theories have had some success with space, time, causality, conservation, and number, while Kohlberg's extension of Piaget's model to the moral sphere has turned out to be a dismal failure, an utterly degenerate research program despite many true believers.

25. Ibid., p. 302.
26. John Dewey, *Human Nature and Conduct* (Carbondale: Southern Illinois University Press, 1988), p. 56.
27. In all likelihood, the capacities for reflection are rooted in the architecture of a system attuned to both similarities and differences, the latter producing cognitive adjustment of various sorts, including proto-reflection.
28. Dewey, *Human Nature*, p. 56.
29. Ibid., p. 58. It is an illusion to which moral theorists are historically, but perhaps not inevitably, prone that their pronouncements can move outside the space of human nature, conduct, and history and judge from the point of view of reason alone what is right and good. This is a silly idea. The key to a successful normative naturalism is to conceive of "reasoning" in the cognitive loop as a natural capacity of a socially situated mind/brain and to argue for the superiority of certain methods of achieving moral knowledge and improving the moral climate. Arguments for the superiority of certain methods and certain norms need to be based on evidence from any source that has relevance to claims about human flourishing.
30. We might also compare the view being floated to Allan Gibbard's norm expressivist view in *Wise Choices*. For Gibbard, norms express judgments about what it makes sense to feel, think, and do. Norm expression involves endorsement, as it does in classical emotivism. But it is endorsement that can be rationally discussed among conversants —endorsement that can be "discursively redeemed," to give it its Habermasian twist (p. 195). Talk of the literal truth or falsehood of normative utterances generally—and normative moral utterances specifically—to one side, there is still plenty to be said on behalf of the rationality or sensibility of particular normative claims. There are conversational challenges to be met, facts to call attention to, consistency and relevance to be displayed, and higher-level norms that govern the acceptance of other norms to be brought on the scene. We will need to explain whether the norms we endorse forbid a particular act, require it, or simply allow it. And we may be called upon to explain the point behind any of our norms.

For Gibbard, morality is first and foremost concerned with interpersonal coordination and social stability. Given linguistic capacity, it is natural that the need for coordination will lead eventually to norms of conversation that govern normative influence, to the vindication of specific normative judgments, and so on. It makes sense to develop such norms if our aim is coordination. We expect our fellow discussants to be consistent in their beliefs and attitudes, and thus we can ask them reasonably to show us how it is that their normative system is consistent. We can ask them to say more about some normative judgment to help us to see its rationale. We can ask them to provide a deeper rationale or a reason for taking them seriously as reliable normative judges, and so on. A "reason" is anything that can be said in favor of some belief, value, or norm—any consideration that weighs in favor of it. Since the things that can be said on behalf of almost anything constitute an indefinitely large class, reasons abound in normative life. No set of reasons may yield conviction in the skeptic or in someone whose life form lies too far from our own. But this in no way distinguishes normative life from the rest of life.

31. I am using "rational," "rationality," and their suite without raised-eyebrow quotation marks from here on, since I have tried to explain that I am using them as they will need to be understood from the point of view of postmodern connectionist mental architecture.

32. This attitude continues to reveal itself in Churchland's forthcoming book, *The Engine of Reason, The Seat of the Soul*, (MIT, 1995).

33. Churchland, *Neurocomputational Perspective*. See also Owen Flanagan, "Quinean Ethics," *Ethics* 93 (1982):56–74.

34. Ibid., p. 300.

35. Rawls, *Political Liberalism*, (New York: Columbia University, 1993), sees the tension as one between the ends of politics and ethics, but the basic point is the same.

36. Stanley Hauerwas has pointed out to me that this example is less plausible empirically than this one: it is wrong to kill someone except when the person is being sacrificed for the sake of the gods.

37. Of course, much of science, even chemistry, biology, and physics, is local. Ecology is local with a vengeance. The generalizations true in tropical and Arctic zones may have no interesting commonalities. The point is that the first few of these aspire to nonlocal generalizations that will be true locally, so if there is any light around, it is traveling at *the* speed of light.

38. Dewey, *Human Nature*, pp. 204–205.

39. We postmodern bourgeois liberals no longer tag our central beliefs and desires as 'necessary' or 'natural' and our peripheral ones as 'contingent' or 'cultural.' This is partly because *anthropologists, novelists, and historians* have done such a good job of exhibiting the contingency of various putative necessities ... [and] philosophers like Quine, Wittgenstein and Derrida have made us wary of the very idea of a necessary-contingent distinction. These philosophers describe human life by the metaphor of a continual reweaving of a web of beliefs and desires." Rorty, "On Ethnocentrism," 1:208.

40. Why think of ethics naturalized as a branch of human ecology? As I insist in *Varieties*, the principle of minimal psychological realism (PMPR) is not sufficient to fix correct theory because many more theories and person-types are realizable than are good *and* because many good ones have yet to be realized. Moral theories and moral personalities are fixed (and largely assessed) in relation to particular environments and ecological niches, which change, overlap, and so on. Therefore, it is best to think of ethics as part of human ecology—as neither a special philosophical discipline nor a part of any particular human science. Are all ways of life okay? Are the only legitimate standards of criticism "internal" ones? The answer is no. What is good depends a great deal on what is good for a particular community, but when that community interacts with other communities, then these get a voice. Furthermore, what can seem like a good practice or ideal can, when all the information from history, anthropology, psychology, philosophy, and literature is brought in, turn out to have been tried, tested, and turned out not to be such a good idea. So if ethics is part human ecology, and I think that it is, the norms governing the evaluation of practices and ideals will have to be as broad as possible. To judge ideals, it will not do simply to look and see whether healthy persons and healthy communities are subserved by them in the here and now. We require that current "health" is bought without incorporating practices—slavery, racism, sexism, and the like—that can go unnoticed for some time and keep persons from flourishing, and eventually poison human relations, if not in the present, at least in nearby generations.

41. It seems to me a common worry about naturalism that it will be conservative. One can see why excessive dependence on the notions of the self-correcting aspects of morality might fuel this worry. This is one reason that I recommend rejecting the self-correcting, progressive idea. It is naively Panglossian.

42. This point about the perspectival nature of moral critique is widely granted across philosophical space. See, for example, all of Alasdair MacIntyre's and Richard Rorty's recent work; C. Taylor, "Understanding and Ethnocentrism," *Philosophical Papers II* (Cambridge: Cambridge University Press, 1985). Hilary Putnam writes, "We can only hope to produce a better conception of morality if we operate from within our own tradition." H. Putnam, *Reason Truth and History* (Cambridge: Cambridge University Press, 1981), p. 216. The point is that there is no other tradition to operate from, no other perspective to take as one begins and engages in the project of moral critique. Of course, the moral source one starts from changes as critique proceeds, and one can modify one's source by adopting the convictions of other individuals or groups.

43. I was continually questioned about this—that which I tried to hold off to another time and place—in both Los Angeles and St. Louis. So for the sake of readers who wonder how the argument would go, here is a sketch that goes with the previous footnote. The overall argument can be correctly read as constituting a defense of some form of normative relativism. The idea that identity, meaning, and morals are all deeply contingent and that our current dilemma is which attitude or attitudes to adopt in the face of this contingency is unabashedly relativistic. The question that invariably arises is how a relativist can make any credible value judgments. A relativist might try to influence others to adopt certain norms, values, and attitudes, but this could not be done rationally, since the relativist does not believe in rationality. This is an old debate, one the antirelativist cannot win. The charges against the relativist are indefensible and invariably unimaginative. We are to imagine that the relativist could have nothing to say about evil—about Hitler, for example. Here I can only gesture toward a two-pronged argument that would need to be spelled out in detail. The first prong involves emphasizing that relativism is the position that certain things are relative to other things. "Being a tall person" is relative. Relative to what? Certainly not to everything. It is relative to the average height of persons. It is not relative to the price of tea in China, or to the number of rats in Paris, or to the temperature at the center of the earth, or to the laws regarding abortion. The relativist is attuned to relations that matter, to relations that have relevance to the matter at hand. Even if there is no such thing as "transcendent rationality" as some philosophers conceive it, there are perfectly reasonable ways of analyzing problems, proposing solutions, and recommending attitudes. This is the essence of pragmatism. Pragmatism is a theory of rationality.

The second prong of the argument involves moving from defense to offense. Here the tactic is to emphasize the contingency of the values we hold dear while at the same time emphasizing that this contingency is no reason for not holding these dear and meaning-constitutive. If it is true, as I think it is, that whether consciousness of the contingency of life undermines confidence, self-respect, and their suite depends on what attitudes one takes toward contingency, then there are some new things to be said in favor of emphasizing "consciousness of contingency." Recognition of contingency has the advantage of being historically, sociologically, anthropologically, and psychologically realistic. Realism is a form of authenticity, and authenticity has much to be said in its favor. Furthermore, recognition of contingency engenders respect for human diversity, which engenders tolerant attitudes. This has generally positive political consequences. Furthermore, respect for human diversity and tolerant attitudes are fully compatible with deploying our critical capacities in judging the quality and worth of alternative ways of being. There are judgments to be made of quality and worth for those who are living a certain way, and there are assessments about whether we should try to adopt certain ways of being that are not at present our own. Attunement to contingency, plural values, and the vast array of possible human personalities opens the way for use of important and underutilized human capacities: capacities for critical reflection, seeking deep understanding of alternative ways of being and living, and deploying our agentic capacities to modify our selves, engage in identity experimentation, and meaning location within the vast space of possibilities that have been and are being tried by our fellows. It is a futile but apparently well-entrenched attitude that one ought to try to discover the single right way to think, live, and be. But there is a great experiment going on. It involves the exploration of multiple alternative possibilities, multifarious ways of livings—some better than others and some positively awful from any reasonable perspective. The main point is that the relativist has an attitude conducive to an appreciation of alternative ways of life and to the patient exploration of how to use this exposure in the distinctively human project of reflective work on the self, on self-improvement. The reflective relativist, the pragmatic pluralist, has the right attitude —right for a world in which profitable communication and politics demand respect and tolerance but in which no one expects a respectful, tolerant person or polity to lose the capacity to identify and resist evil where it exists, and right in terms of the development of our capacities of sympathetic understanding, acuity in judgment, self-modification, and, on occasion, radical transformation.

Chapter 3
How Moral Psychology Changes Moral Theory
Mark L. Johnson

The Moral Philosophy versus Moral Psychology Split

A great many philosophers think that moral philosophy does not have to pay much attention to moral psychology. They think either that moral psychology is mostly irrelevant to moral theory, or they believe that rational self-reflection alone can generate an adequate set of psychological assumptions without relying on any empirical studies from moral psychology. Moral purists of this sort labor under the illusion that there exists a large gulf separating moral theory from moral psychology. They regard "pure" moral philosophy as being concerned only with how we ought to reason and act and with justifying the fundamental principles of morality. They then contrast this sharply with moral psychology, which they allege to be a merely empirical discipline describing the contingent facts about how people actually are motivated, how they understand things, and the factors that affect their moral reasoning. Armed with this grand distinction between moral theory and moral psychology, along with its attendant assumptions of an is-ought split and a fact-value dichotomy, defenders of a very narrow conception of moral philosophy pretend to dismiss the complex, messy concerns of moral psychology, which are regarded as being irrelevant, or at least tangential, to the tasks of moral theory.

Those who want to deny the relevance of moral psychology to moral theory typically try to make their case by assuming an extremely narrow and trivial conception of moral psychology as being concerned only with the psychological conditions that affect concrete deliberations and decisions within specific situations. Knowing why this or that individual or group · reasoned and acted in a certain way, for instance, certainly does not tell whether they acted in a morally praiseworthy manner. Knowing why so many people were attracted to the values and social institutions of nazism does not indeed tell whether those values and institutions were good or bad. Consequently, this trivialized conception of moral psychology can make it seem as though psychology has no important relation to moral theory.

Moral psychology, however, is not psychology in this narrow sense. Rather, moral psychology should be understood broadly as what I will call the psychology of human moral understanding, which includes empirical inquiry into the conceptual systems that underlie moral reasoning. The psychology of moral understanding can give us profound insights into the origin, nature, and structure of our basic moral concepts and into the ways we reason with those concepts. There is no direct deductive link between such knowledge of our moral concepts and specific moral rules (such as rules telling us why nazism is immoral), and that is why moral psychology is not going to give an exhaustive set of prescriptions for moral living. However, moral psychology will tell what is involved in making moral judgments, and it will thereby cultivate in us a certain wisdom that comes from knowing about the nature and limits of human understanding—a wisdom that will help us live morally insightful and sensitive lives.

As a graduate student in the early 1970s, I was indoctrinated, as many other generations of philosophy students have been, with both this trivializing view of moral psychology and also with an extremely restrictive conception of the nature of moral philosophy that has its roots partly in Enlightenment epistemology and partly in the influential pronouncements of G. E. Moore about the nature of ethics. According to the received view, there were alleged to be three radically distinct enterprises that jointly made up the field of ethics:

1. *Descriptive ethics.* This was thought to be a mere empirical investigation of moral standards and practices across times and cultures. As merely descriptive, it was alleged to have no "prescriptive" force, that is, no direct bearing on philosophic attempts to determine how we ought to act.

2. *Normative ethics.* This was supposed to be an attempt to lay down prescriptive moral principles meant to guide our action, our willing, and our moral evaluation of actions and persons.

3. *Metaethics.* This was conceived as a form of conceptual analysis of the cognitive status and semantic content of various moral concepts. Moore's *Principia Ethica* went a long way toward defining moral philosophy as primarily concerned with clarification and analysis of our fundamental moral concepts, such as good, right, duty, and rule. In itself, metaethics was not believed to be a prescriptive or normative activity, although there were intimations that good conceptual analysis would clarify a number of issues relevant to normative ethics.

The nearly exclusive focus on metaethics that characterized at least the first six decades of this century in Anglo-American philosophy was surely the nadir of moral theory. This impoverished state of moral theory explains why so many of us felt considerable excitement and liberation in the late 1960s and early 1970s when we first encountered the work of John Rawls. Rawls sidestepped the dominant metaethical questions and went straight to the genuine normative concerns of moral and political theory. His sophisticated nonfoundationalist epistemology seemed to make it possible for us to do normative ethics once again. However, in spite of this new wave of constructive moral theory, the split between moral philosophy and moral psychology was never seriously questioned. Rawls said only that any adequate moral theory must be generally compatible with our most reliable theories of moral psychology. He did not, however, give a central role to questions of moral psychology, and this has left the moral theory versus moral psychology dichotomy relatively intact for much of the last quarter-century.

The obvious question that arises is why this split between moral theory and moral psychology runs so deep in twentieth-century moral philosophy. The answer, I believe, is tied up inextricably with our traditional conception of the role that moral philosophy ought to play in our lives. People tend overwhelmingly to want their moral philosophy to give them moral guidance for their lives. People want a rational way—a method—for determining how they ought to act in the kinds of situations they typically encounter. A moral theory that does not lead fairly directly to prescriptions for action would be, on this account, no moral theory at all.

This desire for a moral guidance is natural and understandable enough, given the complexity and indeterminacy of our human experience. However, it is a short misstep from this reasonable desire for moral guidance to the claim that any satisfactory moral theory ought to be what I call a "governance" theory, that is, one that gives a set of moral rules specifying how we should act in concrete situations. In contemporary theory, Alan Donagan has articulated a prototypical governance conception, in which morality is defined as "a standard by which systems of mores, actual and possible, were to be judged and by which everybody ought to live, no matter what the mores of his neighbors might be." Moral theory, on this view, is thus "a theory of a system of laws or precepts, binding upon rational creatures as such, the content of which is ascertainable by human reason."[1]

Donagan's theory is quintessentially a governance theory, one that purports to set out definite moral rules that specify how we ought to deal with the complex moral problems of contemporary life. Like Rawls and most other important contemporary moral theorists, Donagan would certainly have thought that no moral theory can be acceptable that presupposes a view of human psychology that is demonstrably false. Almost everyone today would say, at least, that we cannot ignore moral psychology in our moral philosophizing.

Still, many moral philosophers do ignore what is going on in moral psychology, and in the cognitive sciences generally. It is a depressing fact that the moral theory versus moral psychology split still stands, for theoretical reasons or simply because psychology continues to be neglected by philosophers. The impoverished state of twentieth-century moral philosophy is shown by the fact that we so desperately needed a book like Owen Flanagan's *Varieties of Moral Personality*, which devotes 400 pages to defending the proposition that moral theory must incorporate a realistic human psychology. Flanagan has taken the first major steps in setting out the general parameters of an adequate moral theory that would satisfy what he calls the Principle of Minimal Psychological Realism: "Make sure when constructing a moral theory or projecting a moral ideal that the character, decision processing, and behavior prescribed are possible, or are perceived to be possible, for creatures like us."[2] In other words, only recently Flanagan found it necessary to mount a major offensive in order to make the world safe for moral psychology. It is a good and important book, but he should not have been forced to spend so much energy reminding us of the necessity of a psychologically realistic moral theory.

Even more recently, Samuel Scheffler has emphasized "the importance for moral philosophy of some tolerably realistic understanding of human motivational psychology," explaining that he is "convinced that the discussion of some of the central questions of moral philosophy could only benefit from a more serious attention to psychological reality."[3] Unfortunately, Scheffler proceeds to ignore, for the most part, the large body of work in moral psychology that could be relevant to his argument. Feminist moral philosophers have also been arguing for a psychologically realistic moral theory for well over a decade, uncovering a network of assumptions that underlie the dominant conception of moral theory, with all its foundational dichotomies and gendered concepts. And we can, of course, go back to philosophers like James and Dewey for robust moral psychologies that could provide the basis for realistic moral theories.

What I have been urging so far is this: many philosophers cling tenaciously to the moral philosophy versus moral psychology split because they think that this is the only way to preserve a governance theory of morality and, with it, the idea that moral philosophy can give us moral guidance. They fear that wading into moral psychology can only teach us how and why people do what they do, but without telling us definitively what people ought to be doing.

I am going to argue that the central purpose of moral theory should be the enrichment and cultivation of moral understanding and that moral psychology is essential to the development of our moral understanding. Moral psychology therefore lies at the heart of any adequate moral theory. There will be moral guidance from moral theory so construed, but only of the sort that comes from moral insight into complex situations and personalities, rather than the sort that comes from applications of moral rules. This is all the guidance we can have, all that we have ever had, and all that we need.

Why Do We Need to Incorporate Moral Psychology?

The answer to the question of why moral theory needs a robust moral psychology is this: our morality is a human morality, and it must thus be a morality directed to our human concerns, realizable by human creatures like ourselves, and applicable to the kinds of problematic situations we encounter in our lives. This means that we cannot do good moral theory without knowing a tremendous amount about human motivation, the nature of the self, the nature of human concepts, how our reason works, how we are socially constituted, and a host of other facts about who we are and how the mind operates. Moreover, we cannot know how best to act unless · we know something about the details of mental activity, such as how concepts are formed, what their structure is, what constrains our inferences, what limits there are on how we understand a given situation, how we frame moral problems, and so forth. Without knowledge of this sort, we are condemned to either a fool's or a tyrant's morality. We will be fools insofar as we make stupid mistakes because we lack knowledge of the mind, motivation, meaning, communication, and so forth. Or we will suffer the tyrannical morality of absolute standards that we impose on ourselves and others, without any attention to whether people could actually live up to such standards, apply them to real situations, and improve life by means of them.

Over seventy years ago, John Dewey made the case for the empirical character of moral theory, arguing that because morality involves deliberation about possible courses of action, a vast range of empirical knowledge about action, desire, and reasoning is centrally relevant to moral philosophy:

> But in fact morals is the most humane of all subjects. It is that which ·
> is closest to human nature; it is ineradicably empirical, not theological nor metaphysical nor mathematical. Since it directly concerns human nature, everything that can be known of the human mind and body in physiology, medicine, anthropology, and psychology is pertinent to moral inquiry.... Hence physics, chemistry, history, statistics, engineering science, are a part of disciplined moral knowledge sofar as they enable us to understand the conditions and agencies through which man lives, and on account of which he forms and executes his plans. Moral science is not something with a separate

province. It is physical, biological and historic knowledge placed in a human context where it will illuminate and guide the activities of men.[4]

I will argue that the exclusion from moral philosophy of a robust moral psychology stands directly in the way of genuine moral understanding and insight. Once we challenge this foundational dichotomy, it becomes necessary to reevaluate the nature and purpose of moral philosophy. The bottom line is that moral philosophy should be a theory of moral understanding, which necessarily incorporates the empirical results coming from studies in moral psychology and the cognitive sciences.

A comprehensive moral psychology would include at least the following types of inquiry:

1. *Personal identity*. What are the cognitive, affective, and social dimensions of the formative process by which a person develops an evolving sense of self-identity?

2. *Human ends and motivation*. What is the structure of motivation by which goals and purposes develop? Where do our "ends" come from? What gives rise to our conception of various goods?

3. *Moral development*. Are there stages through which people normally pass as they develop what we regard as a mature moral consciousness? Are these stages universal, or do they vary according to race, gender, or cultural differences?

4. *Conceptualization*. What is the semantic structure of human concepts? Where do our concepts come from, and how are they extended to cover new experiences? Are concepts defined by necessary and sufficient conditions, or do they have a more open, variegated, and imaginative internal structure?

5. *Reasoning*. What is the nature of moral deliberation? Is it deductive, inductive, or perhaps an imaginative exploration of possibilities for concrete action? Is it constrained in any way, or is it radically subjective and relativistic?

6. *Affect*. What is emotion? Is affect separable from conceptualization, reflection, and reasoning, or is it inextricably woven into the fabric of all experience? How is affect related to motivation? Is there cognitive and inferential structure to the emotional dimensions of experience?

This is only a partial list of the basic areas within moral psychology and the cognitive sciences that must be considered part of any adequate moral theory. Most people do not think of cognitive science as having any bearing on morality. They hold this prejudice for two basic reasons. First, they are biased against the "empirical," since they hold some version of the fact versus value dichotomy. Second, they have an extremely narrow conception of cognitive science as formalist, reductionist, and inhumane.

But the fact is that the cognitive sciences have evolved considerably in the last decade. A second generation of cognitive science has emerged that is neither reductionist nor overly formalist. Traditional, first-generation cognitive science was defined by artificial intelligence, information processing psychology, generative linguistics, and formal model theory. It took the MIND AS COMPUTER PROGRAM metaphor quite seriously, and it had (and has) virtually nothing to say about morality, politics, social theory, and social relations. By contrast, the newly emerging second generation of cognitive science recognizes the embodied and imaginative character of all human conceptualization and reasoning. It focuses on the social, interactive, evolutionary character of human experience and understanding. It looks at human cognition as embodied in a growing biological organism that is interacting and co-evolving with its physical, social, and moral environments. Cognitive science of this sort has plenty to say about morality, and it has a major contribution to make to moral understanding.

It is obviously impossible to survey the full range of relevant empirical results from the cognitive sciences that bear directly on moral theory. I propose to consider just one small part of the new discoveries we are making in second-generation cognitive science that change our view of what moral theory is. In particular, I call attention to the metaphoric nature of our most basic moral concepts and then ask whether this fact requires us to reassess both our understanding of moral experience and our conception of moral theory.

The Metaphoric Nature of Moral Understanding

In the past several years, one of the most robust and potentially revolutionary findings about the mind has been the discovery that the human conceptual system is fundamentally and irreducibly metaphoric. A large and rapidly growing number of studies have shown that our basic concepts in virtually every aspect of human experience are defined by systems of metaphors.[5] A conceptual metaphor is a mapping of conceptual structure from a source domain, which is typically some aspect of our concrete bodily experience, onto a more abstract or less highly articulated target domain. It is crucial to keep in mind that conceptual metaphors are conceptual. They are structures in our conceptual system, not merely propositions or linguistic entities. They involve conceptual structure, the basis for the inferences we draw from the metaphor. The content and logic of the source domain thus determines our understanding of the target domain. In other words, the reasoning we do about the target domain is based on the embodied corporeal logic of the source domain. In this way, our systematic conceptual metaphors do not merely highlight preexisting structures in two different domains; rather, the structure and knowledge pertaining to the source domain partly construct our knowledge in the target domain.

As an example of the value and importance of moral psychology for moral theory, I focus on some recent work on the metaphorical nature of our basic moral concepts.[6] We are just beginning to examine the complex web of systematic metaphors by means of which we define our values, ends, actions, principles, and every other aspect of our moral experience. Moreover, because our moral concepts are defined by systems of metaphors, our moral reasoning is based on the logic of these metaphors.

An incident in Amy Tan's *The Joy Luck Club* provides a concrete example of what I mean by metaphors of morality. Ying-ying was a beautiful, refined young woman living a luxurious and carefree life with her wealthy family just before World War II. "When I was a young girl in Wushi," she remembers. "I was *lihai*. Wild and stubborn. I wore a smirk on my face. Too good to listen. I was small and pretty. I had tiny feet which made me very vain.... I often unravelled my hair and wore it loose." At sixteen she finds herself inexplicably attracted to an older man from another town. Within six months she is married and then realizes that she has actually come to love him.

No sooner is she married than she realizes that he is a womanizing drunkard. He abuses her emotionally, impregnates her, pursues a series of extramarital affairs, and eventually abandons her for an opera singer. This public infidelity humiliates, shames, and destroys her. He has taken her soul and left her a mere ghost:

> So I will tell Lena of my shame. That I was rich and pretty. I was too good for any one man. That I became abandoned goods. I will tell her that at eighteen the prettiness drained from my cheeks. That I thought of throwing myself in the lake like the other ladies of shame. And I will tell her of the baby I killed because I came to hate this man.
>
> I took this baby from my womb before it could be born. This was not a bad thing to do in China back then, to kill a baby before it is born. But even then, I thought it was bad, because my body flowed with terrible revenge as the juices of this man's firstborn son poured from me.

How are we to understand the logic by which this tortured innocent comes to kill her baby? In her mind it is an act of revenge. But what is the logic of revenge? In brief, Ying-ying's husband has taken her most precious possession: her spirit, her *chi*, her honor. She has "lost face" and is in shame. Ying-ying exacts her revenge by taking the most precious possession she can from him: his firstborn son: "When the nurses asked what they should do with the lifeless baby, I hurled a newspaper at them and said to wrap it like a fish and throw it in the lake." She symbolically drowns the baby in the lake, just as the women of shame drown themselves in the lake.

The logic of this tragic action stems from what I have called elsewhere the MORAL ACCOUNTING metaphor, which is concerned primarily with what we owe to other people to increase their well-being and what they, in turn, owe to us. Basically, we understand our moral interactions metaphorically as a species of economic transaction, according to the following conceptual mapping:

The MORAL ACCOUNTING *Metaphor*

Commodity Transaction	Moral Interaction
Objects, commodities	Deeds (actions), states
Utility or value of objects	Moral worth of actions
Wealth	Well-being
Accumulation of goods	Increase in well-being
Causing increase in goods	Moral = causing increase in well-being
Causing decrease in goods	Immoral = causing decrease in well-being
Giving/taking money	Performing moral/immoral deeds
Account of transactions	Moral account
Balance of account	Moral balance of deeds
Debt	Moral debt = owing something good to another
Credit	Moral credit = others owe you something good
Fair exchange/payment	Justice

This conceptual mapping provides an experiential basis for a large number of inferences that we draw in evaluating ethical conduct. We use our basic knowledge of the source domain (economic transactions) to make moral inferences about situations in the target domain of moral interactions. Consider, for example, our knowledge about wealth and how it generates inferences about morality via the WELL-BEING IS WEALTH metaphor. Wealth is something that one amasses by owning property (land, commodities) or its surrogate, money. Typically, wealth is a product of labor, which is to say that people earn it by their work, although it may come to them by other means, such as inheritance. Being wealthy usually makes it possible for people to have more of the things they want to have and to do more of the things they want to do. It allows them to satisfy their needs and desires, and it may enhance the quality of their existence. There is a limited amount of wealth available in the world, and it must be divided up among many people. Fair exchange gives each person what is due him or her.

In the context of the MORAL ACCOUNTING metaphor, we thus come to understand moral well-being as wealth, according to the following mapping:

The WELL-BEING IS WEALTH Metaphor

Financial Domain	Moral Domain
Wealth	Well-being
Accumulation of goods	Increase in well-being
Profitable = causing increase in wealth	Moral = causing increase in well-being
Unprofitable = causing decrease in wealth	Immoral = causing decrease in well-being

The WEALTH metaphor is one of the two or three most important conceptions of moral well-being that we have. It shows itself in the ways we think and talk about well-being—for example:

> She has had an undeservedly *rich* life.
> The cynics of the world lead *impoverished* lives.
> Doing disaster relief work has *enriched* Sarah's life immeasurably.
> Prince Charles *profited* from his relationship with Princess Di. He is certainly a better person now.
> I've had a *wealth* of happiness in my life.
> Nothing can compare to the *riches* of family, friends, and loved ones.

Within the MORAL ACCOUNTING metaphor, the WELL-BEING IS WEALTH metaphor gives rise to definite inferences about our moral obligations. That is, we reason on the basis of the metaphor. Given the conceptual mapping from source to target domain and based on our knowledge of the source domain, we then develop a corresponding knowledge in the target domain and draw the appropriate inferences. Moral well-being comes to a person as a result of his or her own efforts and also as something given by the good actions of other people. Moral well-being is something that can accumulate and that can also diminish. The more well-being you have, the better off you are. Immoral action decreases well-being. Consequently, moral acts toward others (acts that increase their well-being) put them in moral debt to you and thereby give you moral credit; you deserve an equal amount of well-being in return for what you have given.

The MORAL ACCOUNTING metaphor thus gives rise to a pattern of reasoning about our duties, rights, and obligations. On the basis of this metaphor, we reason about what is fair, and our moral discourse reveals this underlying conceptual metaphor system. If you do something to diminish my well-being, then you incur a debt to me, morally speaking, since you are expected to "pay me back" for what you have taken. When you perform noble acts, you build up moral credit. Thus, we say such things as: .

In *return* for our kindness, she *gave us* nothing but grief.
In judging him, *take into account* all the good things he has done.
I'm holding you *accountable* for her suffering.
When you compare his kindness with what he is accused of doing, it just *doesn't add up*.
All her sacrifices for others surely *balance out* the bad things she did.
His noble deeds *far outweigh* his sins.
Mary certainly *deserves credit* for her exemplary acts.
I *owe* you my life!
I couldn't possibly *repay* your kindness.
Milken *owes a great debt* to society for his evil doings.
You must *pay* for your selfishness.

Elsewhere I have shown how the MORAL ACCOUNTING metaphor gives rise to a set of at least five basic schemas that people use to evaluate the moral merit of various actions and to determine what is due them, as well as what they owe others. Sarah Taub has outlined schemas for reciprocation, retribution, revenge, restitution, and altruisim by which we draw inferences from the MORAL ACCOUNTING metaphor in deciding who gets moral credit. Take, for example, the REVENGE schema. Let us say that you do something bad to me and thereby diminish my well-being. In this sense, you have taken something from me—some of my well-being— and, via MORAL ACCOUNTING, you now owe me something that will increase my well-being to compensate for what you have taken away. But you will not give me back the measure of well-being that you owe me. Therefore, *I* balance the moral well-being books by taking an equal measure of *your* well-being, thus diminishing your moral wealth. That is why we speak of taking revenge on someone; we are taking something good from the person. The REVENGE schema thus has the following structure:

The REVENGE *Schema*
Event: A gives (does) *something bad* to B.
Judgment: A owes *something good* to B.

Complication: A will not give *something good* to B.
Expectation: B should take *something good* from A.
Moral inferences: A has an obligation to give *something good* to B.
B has a right to receive *something good* from A.
Monetary inference: B exacts payment from A.

For example:

Revenge is "an eye for an eye."
Carry *took revenge* on her classmates.
"I'll *make you pay* for what you did!"
"I'll *take it out of your hide*."
"He'll *get even* with you for this."
"Jane *owes you one* for that." (What she owes is something bad that diminishes your well-being.)

We are now in a position to see why Ying-ying does what she does. We can understand the logic of her reasoning that is based on the REVENGE schema for interpreting the MORAL ACCOUNTING metaphor. In addition to the formal structure of the MORAL ACCOUNTING metaphor, she uses two additional metaphors:

1. FACE (HONOR) IS A VALUABLE POSSESSION.
2. A CHILD IS A VALUABLE POSSESSION.

Ying-ying's husband has taken her spirit. She has lost face and is shamed. She takes away the spirit of his firstborn son. It is an empty revenge that leaves her a ghost floating through time:

I became like the ladies of the lake. I threw white clothes over the mirrors in my bedroom so I did not have to see my grief. I lost my strength, so I could not even lift my hands to place pins in my hair. And then I floated like a dead leaf on the water until I drifted out of my mother-in-law's house and back to my family home.

The REVENGE schema and the other schemas Taub has identified are all modes of expectation, evaluation, and inference that follow from the various ways in which the MORAL ACCOUNTING metaphor can be filled in by various conceptions of well-being and differing kinds of actions. They are constitutive of a large part of the moral reasoning we do when we are trying to decide what to expect from others and how we ought to treat them.

The *Joy Luck Club* example reveals importantly that there are at least two levels of conceptual metaphor operative in our moral judgments. The first ("higher") level consists of metaphors for our moral interactions generally, such as the MORAL ACCOUNTING metaphor, which sets the parameters of our judgments about what is due us and what we owe others. Metaphors at this level define our moral framework and fundamental moral concepts. But in order for our moral frameworks to be applied to concrete situations, we need a second level of metaphor for conceptualizing the situations. The REVENGE schema, for example, is empty without the metaphors of BABY AS VALUABLE POSSESSION and FACE AS VALUABLE POSSESSION that give content to the schema and make it applicable to Ying-ying's situation. In sum, our moral reasoning typically depends on which of several possible metaphors we use at these two basic levels: (1) adopting a particular metaphorically defined framework for our interactions and (2) filling that framework in with metaphors that connect it to the particular situation (such as whether we understand the BODY AS A VALUABLE POSSESSION). These two levels must fit together to give concrete moral inferences. It follows also that moral critique can be directed at either or both of these levels, since we can criticize both our general moral framework and the more specific metaphors we use to understand aspects of situations.

Basic Metaphors for Morality

MORAL ACCOUNTING is only one of several fundamental metaphors that define our moral understanding and reasoning at this first, higher level. So far, we have discovered a small number of other basic metaphor systems for various parts of our conception of morality. Although this list is by no means complete, it does set out some of the most fundamental moral concepts and shows how we reason from them. Here is the list of metaphors as we currently understand them:

1. MORAL ACCOUNTING. Our good deeds increase the moral well-being of others (via WELL-BEING IS WEALTH). They earn us moral *credit*. We *owe* others for the good things they have done for us. We should *repay* their kindness. Our evil deeds create a *debt* to other people and society in general.

Inference patterns: Wealth is a valuable commodity that can be amassed, earned, wasted, stolen, given away according to standards of fair exchange. Therefore, our moral interactions are regarded as modes of moral exchange in which well-being is amassed and lost and in which people build up moral credit and create moral debts through their actions.

2. MORALITY IS HEALTH/IMMORALITY IS SICKNESS. Health requires cleanliness, exercise, proper diet, and rest. When moral well-being is understood as health, it follows that all forms of moral sickness are bad. Bad deeds are *sick*. Moral *pollution* makes the soul sick. We must strive for *purity* by avoiding *dirty deeds*, moral *filth, corruption,* and *infection* from immoral people.

Inference patterns: Sick people spread disease, so we try to stay away from them and to maintain the cleanliness and bodily conditions that allow us to resist disease. Therefore, if moral evil is a disease, we must quarantine those who are immoral (by censoring them and shunning their company), so that we are not exposed to their influence. We must keep ourselves clean, pure, and protected from moral infection.

4. BEING MORAL IS BEING UPRIGHT. When we are healthy and strong, we stand upright against disease and natural forces that might knock us off our feet or lay us low. We have power and control. This same logic applies to being morally *upright*. "The Fall" (into sin and wickedness) is caused by the *force* of evil.

Inference patterns: Natural forces can knock you off your feet and make you lose control, thereby being unable to function successfully. Moral evil is an *ever-present* force that will cause you to lose self-control and the ability to do what is right. Morality is thus a struggle to maintain your moral uprightness, balance, and control.

There are two fundamental dimensions of being morally upright:

4A. MORAL STRENGTH. When you are morally *strong*, you have control over your *lower* self (your desires, passions, and emotions). Morality is a *struggle* between the *higher* moral self and the *forces* of our bodily selves. *Willpower* is essential to maintaining the proper *control* of our passions and *baser* desires.

Inference patterns: Staying in control requires strength to manage the natural forces acting upon you. In moral control, the rational will must be strong if it is to manage the powerful forces of desire and passion that lead you to pursue your animal instincts and wants.

4B. MORAL BALANCE. When you are balanced in an upright posture, you cannot be easily knocked over. When you are internally balanced, every organ works together for health and well-being. Moral *balance* is essential for moral *health*. Each part of a person must perform its proper moral function, lest he or she *fall* into evil ways.

Inference patterns: Keeping a balanced posture is essential for dealing with natural forces that would upset your controlled functioning in your environment. Moral balance therefore is necessary if you are going to maintain control over your ability to act as moral reason requires.

5. BEING MORAL IS BEING IN THE NORMAL PLACE. According to a pervasive metaphor for human action, which I have named the EVENT STRUCTURE metaphor, actions are self-propelled motions along paths to destinations (goals). Certain ends (destinations) are moral *ends*. They are ends we have a duty to realize through our actions. The moral path is *straight and narrow*. Moral *deviance* is a *straying* from the true path. *Violating other people's boundaries* is immoral.

Inference patterns: Motion along a path gets you to your desired destination. In purposeful action, ends or goals are metaphoric destinations on these motion-paths. Moral "ends" are the metaphorical places we should strive to reach, and they are socially, religiously, and morally established. Being moral is going where you ought to go, along paths set up by society. Deviance is immoral because it takes you away from or out of the region you ought to be in, along with other people. Deviants can lead other people astray, and so they are perceived as a threat to the moral community.

6. MORALITY IS OBEDIENCE. Our parents lay down for us rules of acceptable behavior, and they enforce those rules with sanctions, such as punishments and rewards. Moral *obedience* requires that we *follow rules* (obey moral laws) that tell us how to act. Reason *issues* moral *commandments* that we are obliged to obey.

Inference patterns: It is assumed that our parents (as authorities) have our best interests at heart and act for our benefit and that they also know what is best for us as children. Culturally, then, moral authorities are designated people who are supposed to know what is best for you and who act in the interest of your moral well-being. Obeying these authorities is the morally correct thing to do. The moral authorities can be actual people (the *pope*, religious leaders, people who are wise), or they can be metaphorical personifications (Reason, Law, the State).

7. MORAL ORDER. Having things in their proper order (in their place) is necessary for successful functioning. Society is a system—a metaphoric machine, person, building, organism—in which everything must be *in order*, if we are to live and flourish. That is why we need *law and order*. Moral *chaos* threatens to destroy society, causing widespread *breakdown, disintegration,* and *malfunction*.

Inference patterns: For every kind of system, mechanistic or organic, things must be "in order" if the system is to function properly. Things must be in working order. Morally, if everything is not "in order," society will break down and cease to function. Therefore, any disorder or chaotic activity is perceived as a serious threat to society and to human well-being.

8. MORALITY IS LIGHT/IMMORALITY IS DARKNESS. It is scary and dangerous in the dark. One can neither see nor function well in the dark. We tend to regard darkness as the harbinger of bad happenings. Moral *darkness* is a threat to our basic well-being. The *dark side* threatens to overcome the *light* in us. Evil is a *dark force*, and the *Prince of Darkness* is the most evil being of all.

Inference patterns: When it is dark, you cannot see things. You stumble around, lose your way, and cannot function efficiently. Metaphorically, UNDERSTANDING IS SEEING, and so the darkness of evil makes you incapable of seeing the good and knowing what is right and wrong. You lose your moral bearings and stumble about, not being able to get where you ought to be going (to your moral ends). Moral darkness brings ill-being to us.

9. MORAL PROJECTION. Human perception is ineliminably perspectival. We always experience any object from a particular point of view, and the more perspectives we can take up on the object, the more objective our knowledge is considered to be. We rise above some of the limitations of our own way of seeing things by "seeing things as others see them." Morally, projecting ourselves into another person's way of experiencing things helps us rise above our own prejudices in order to act more humanely.

Inference patterns: If you want your perceptual knowledge to be as objective as possible, you try to achieve as many perspectives as possible. If you want objective moral knowledge, you must take up the *moral point of view*—the point of view of a "moral person in general." This means that you must be able to abstract from your prejudices and take up the standpoint of an ideal moral judge who decides how to act on grounds that hold for every person, not just for this or that particular person. Moral empathy is a form of putting yourself in the place of another. All moral theories founded on the basis of a rational moral agent require the metaphor of the projected self (for example, Rawls, ideal-observer, and universal standpoint theories).

This list of basic metaphors for morality is partial and needs extensive further analysis, for example, setting out the structure of the source domain for each metaphor, explaining why that particular source domain is used, laying out the conceptual mapping from source to target domain, and showing how this mapping constrains our moral reasoning. However, even this brief list suggests two very significant points about the metaphoric nature of our moral understanding. First, it is vital to notice that basic moral terms like "ought" and "should" really have meaning and lead to moral inferences only through one or more of the above metaphors, along with the second-level metaphors by which we conceptualize actual situations. "Ought" means one thing and supports certain very specific moral inferences in the context of the MORAL ACCOUNTING metaphor, compared to the quite different set of inferences that it generates relative to the MORALITY IS HEALTH metaphor. According to the MORAL ACCOUNTING metaphor, for instance, "ought" is spelled out in terms of economic transactions of fair exchange, credit, and debt. MORALITY IS HEALTH, by contrast, establishes imperatives that direct us to fight moral sickness and promote certain states of moral flourishing within individuals and the community. "Ought" therefore gets its content and concrete applicability by means of its role in metaphorically defined moral frameworks.[7]

Second, it is remarkable that there seem to be so few basic source domains for our metaphors of morality. Why should we have these source domains and not others? My research so far suggests the following general answer. First, these source domains appear to be universal in human experience because they depend on the nature of our bodies and their typical interactions with the types of environments we inhabit. Second, these source domains are characteristically tied up with our sense of personal and communal well-being, growth, and satisfaction, which makes them suitable source domains for metaphors of morality. It is no accident that source domains such as health, balance, strength, movement to a place, obedience, and light/dark are intimately tied up with our sense of human flourishing, and so they are prime candidates for universal metaphors of morality. Whether these are, in fact, universal source domains is a matter for empirical study, but there is some evidence to think that they may be found in all cultures, even if they are not elaborated in just the same way in each culture.

Metaphor and Moral Reasoning

The most important epistemological and moral implication of the fact that our basic moral concepts are defined by metaphors is that we reason on the basis of these metaphors about how we ought to act and what kind of person we ought to be. The logic of the source domain, as it is mapped onto the target domain, constrains the inferences we make about the target domain of morality. We have seen this already in the way that the REVENGE schema leads to judgments and actions within the framework of the MORAL ACCOUNTING metaphor. The crucial point is that each metaphor has its own logic and generates epistemic entailments about the target domain (which is here some part of morality).

In order to show how strongly these metaphors constrain our moral reasoning, let us consider the metaphor MORALITY IS HEALTH. The relevant conceptual mapping is:

The MORALITY IS HEALTH Metaphor

Physical Health	*Moral Behavior*
Health	Well-being
Sickness	Moral degeneration
Pollution	Cause of evil
Being diseased	Being morally depraved
Physical exercise	Moral training
Growth	Moral improvement

Notice the very strong way in which the logic of the source domain (physical health) determines the logic of the target domain and constrains our reasoning. What is it that makes you unhealthy? The answer is, any cause of disease—infection, plague, pollution, filth, and, in general, things that you find disgusting. What promotes health? The answer is, exercise, watching what you eat, avoiding those who are sick, staying clean, and so forth. Knowledge of this sort from the source domain carries directly over into our reasoning about morality. We speak of moral and social *diseases*, of *sick* acts, of *dirty* pictures that *pollute* people's minds, of *filth* and *smut*, of *disgusting* behavior, and *cancer* at the heart of society. Sick people can spread disease. Consequently, in order to avoid becoming sick ourselves, we stay away from them, quarantine them, try to kill the disease, and try to keep ourselves clean. Morally, then, we argue that immoral people can *infect* others with their evil. We do everything we can to distance ourselves and our children from the causes of immorality and the people we regard as immoral. We try to *clean up* our schools and our towns. We treat illegal drugs as a *plague* on society. We think that association and contact with someone who does an immoral act can cause a *diseased* mind.

In short, the *logic* of the metaphor determines our expectations, our reasoning, and our actions. The HEALTH metaphor is evidenced in a huge variety of expressions we use to make moral judgments—for example:

His intentions were *pure*, even if things didn't work out well.
She has no moral *blemishes*.
"... without *spot* of sin."
"O Lord, create a *pure* heart within me."
"What a *stinking, low-down, dirty* trick, you miserable rat!"
Scarlett was *washed clean* of her sin.
Pornography *pollutes* the mind and soul.
We must keep that *filth* out of our schools.
Crime is a *plague* that *infects* us all.
He lives in a cocaine *sewer*.
What a *sick, disgusting* thing to do.

These conceptual metaphors are not merely optional ways of talking about morality. There is nothing optional about them at all, and they are not merely matters of words. They are the means by which we define our moral concepts. Although we may not be limited to just one unique metaphor system for a particular concept (for example, we have both WELL-BEING IS WEALTH and WELL-BEING IS HEALTH), neither can we use just any metaphor, especially since this is seldom a matter of conscious choice and the range of metaphors available is relatively small. Most important, the nature and structure of the source domain constrains the inferences we make about the aspect of morality that is the target domain. The metaphor, in other words, sets limits on our reasoning about how we ought to behave and how we ought to regard others. If, for instance, we understand moral well-being as WEALTH, we will act and reason quite differently than if we understood it as HEALTH. The logic of the WEALTH metaphor contains notions of fair exchange, quantification, and balance, while the logic of moral HEALTH emphasizes avoidance of immoral people, staying pure, fighting moral disease, and maintaining moral discipline.

How Cognitive Science Changes Ethics

Having surveyed some of the basic metaphors that define our moral understanding, the nagging question arises, So what? "So what?" the moral apriorist will ask. "What difference could it possibly make to learn that people typically use metaphors to understand their experience? We want to know how they *ought* to reason, not how they tend to reason."

The answer to this question is clear and straightforward, and in offering an answer, I am suggesting in general how empirical studies in the cognitive sciences bear on morality and moral theory. The general answer is that our morality is a *human* morality, one that must work for people who understand, and think, and act as we do. Consequently, if moral theory is to be more than a meaningless exploration of utopian ideals, it must be grounded in human psychology.

The moral purist, in pursuit of the illusory ideal of a strict governance theory of morality, demands a nonexistent direct connection between moral understanding and morally correct action. The only answer that a moral purist will allow is one that shows how learning about the metaphoric structure of morality, for instance, would lead, in a step-by-step fashion, directly to rules that would tell us how to act. But this is not possible in any but the most obvious, well-worn, unproblematic cases. Knowing the nature and entailments of the MORAL ACCOUNTING metaphor does not tell us whether MORAL ACCOUNTING is a good form of moral interaction in any particular situation. However, knowing all we can about the MORAL ACCOUNTING metaphor can help us make informed judgments about the probable consequences of acting on the basis of this particular metaphor.

It is extremely important to see that the moral purist's charge that cognitive science and moral psychology have no direct bearing on moral theory is based on an illusory fact-value dichotomy that manifests itself in two main fallacies:

1. *The Independence of Facts fallacy:* Facts are independent of any value assigned to them.
2. *The Irrelevancy of the Empirical fallacy:* Facts do not tell you what value to assign to them.

The Independence thesis has been demolished most prominently by arguments in the philosophy of science showing that whatever is counted as a "fact" depends on certain values we have, such as our interests, purposes, criteria of importance, or models. The Irrelevancy thesis is more difficult to falsify because it contains the kernel of truth that if we understand moral psychology as merely a description of why this or that person or community reasons or acts in a certain way, then indeed this will not tell you how they ought to reason and act. These kinds of descriptions do not supply a basis for moral critique.

Nevertheless, the Irrelevancy thesis can be shown to be false once we introduce a much more profound sense of moral psychology and cognitive science that does have normative implications for our moral judgments. As it applies to moral theory, cognitive science in this richer sense is the empirical study of how we conceptualize values and reason about them. This aspect of moral psychology *does* have normative and critical implications, because it gives us insight into the nature of our values and how they constrain our inferences about moral matters. For example, the empirical study of our moral conceptual system reveals the metaphors that define our moral frameworks, and it can open our eyes to the limitations of this or that metaphor of morality. It can show us what our metaphors highlight and what they hide. It reveals the partial nature of any metaphorical conceptualization and of the reasoning we do based on each metaphor, and it shows us that we may need multiple conceptualizations to discern the full range of possibilities open to us in a given situation. Knowledge of this sort is knowledge that should influence our judgments and actions. It is knowledge that comes from what I earlier called the psychology of moral understanding, which I contrasted with trivialized moral psychology, that is, moral psychology with blinders. A rich psychology of moral understanding looks not merely at people's beliefs and motivations but especially at their deepest moral concepts and the reasoning that stems from them.

The absence of rules for deriving moral judgments from knowledge about metaphorical concepts is not something to be lamented. It is simply a fact about the complexity of human moral understanding, and it is an extremely important fact that has the following significant implications for ethics.

Conceptual Analysis
If our basic moral concepts are metaphoric, then conceptual analysis must presuppose some view of the nature of metaphor as underlying the analyses. Any moral theory that does not recognize the metaphoric nature of moral concepts must be inadequate, and probably disastrously so. Furthermore, the adequacy of the theory will depend on the adequacy of its

theory of metaphor. If we are going to get insight into morality, we need a view of metaphor that recognizes its central role in understanding and reasoning. Whatever remains of "metaethics," therefore, is to a significant extent an exercise in metaphor analysis.

Moral Reasoning
If our basic moral concepts are metaphoric and if we use metaphors to frame the situations we are deliberating about, then our moral reasoning is primarily an exploration of the entailments of the metaphors we live by. For the most part, then, moral reasoning is not deductive, and it is not primarily a matter of applying universal moral principles or rules to concrete situations. I have shown why this model cannot work for the kinds of beings we are and for the kinds of situations we encounter. The reason is that the traditional deductive model has no place for metaphoric concepts, or for any concepts that do not have classical (that is, necessary and sufficent conditions) structure.[8]

Partial Understanding
It follows from the imaginative nature of moral concepts and reasoning that no understanding is exhaustive or comprehensive. Human moral understanding is a complex cluster of metaphor systems, some of which are mutually inconsistent, and yet we manage to live with them and plot our lives by them.

Beyond Absolutism
Because our moral understanding is necessarily partial, morality is not a set of absolute, universal rules but an ongoing experimental process. We must continually be experimenting with new possibilities for action, new conceptions of human flourishing, and new forms of interaction that permit us to adjust to, and also to manage, the ever-changing conditions of human existence. As long as we and our entire ecological situation are evolving, morality must remain experimental. Any attempt to codify this procedure into a final method or absolute principles is a recipe for moral rigidity and obtuseness.

Grounded Moral Theory
The partial, nonabsolute character of our moral understanding might make one think that morality is historically and culturally contingent in a radical way. It would then seem that there is no point in trying to construct a normative moral theory. This is a mistaken view. The most basic source domains for our metaphors for morality are grounded in the nature of our bodily experiences and tied to the kinds of experiences that make it possible for us to survive and flourish, first as infants and then as developing

moral agents. Whether these basic source domains are universal is an open question that awaits further cross-cultural investigation. But if anything is universal, we have good reason to think that structures such as these will be. I believe that these experiential source domains provide general constraints on what can be a psychologically realistic morality, as well as an adequate moral theory. The general nature of such constraints suggests, as Owen Flanagan has argued at length, that there will always be a plurality of appropriate conceptions of human flourishing and a range of possible ways of realizing such conceptions of the good. Although these constraints do not underwrite a universal governance theory, they do limit the range of acceptable alternatives.

Moral Imagination

Moral deliberation is an imaginative enterprise in which we explore the possibilities for enhancing the quality of human existence in the face of current and anticipated conditions and problems. When we are trying to figure out the best thing to do in a given situation, we are tracing out the implications of various metaphors to see what they entail concerning how we should act. Projecting possible actions to determine their probable results, taking up the part of other people who may be affected, and reading with sensitivity the relevant dimensions of a particular situation are all forms of imaginative activity.

People who stress the imaginative and affective dimensions of human understanding are often mistakenly accused of being irrationalist and subjectivist. This serious misinterpretation is the result of a continued adherence to traditional rigid distinctions between such capacities as perception, imagination, feeling, and reason. Stressing the imaginative nature of our moral understanding in no way impugns the rationality of morality. I am arguing here for an enriched conception of human reason as fundamentally imaginative. My point is that moral reasoning is a much richer, more complex, and more flexible capacity than it has been conceived to be in traditional Enlightenment accounts of practical reason. Moral reasoning *is* reasoning, but it is a reasoning that is thoroughly imaginative in character.

What Should a Theory of Morality Be?

A theory of morality should be a theory of moral understanding. Its goal should be moral insight and the guidance and direction that come from a deep and rich understanding of oneself, other people, and the complexities of human existence. At the heart of moral reasoning is our capacity to frame and to realize more comprehensive and inclusive ends that make it possible for us to live well together with others. It involves an expansive form of imaginative reason that is flexible enough to manage our changing experience and to meet new contingencies intelligently. The key to moral intelligence is to grasp imaginatively the possibilities for action in a given situation and to discern which one is most likely to enhance meaning and well-being.

The idea of moral theory as providing governance through rules and principles is fundamentally mistaken. In fact, it is counterproductive to the extent that it overlooks the changing character of experience and does not allow us to see creatively new possibilities for action and response. As Dewey saw, what moral principles we have are not technical rules but rather "empirical generalizations from the ways in which previous judgments of conduct have practically worked out."[9] They are summaries of strategies that have proved more or less useful for the kinds of situations previously encountered. But they must never be allowed to solidify into absolute rules, for then the opportunity for moral growth and progress is undermined.

It should now be obvious why I think that the alleged split between moral theory and moral psychology is not just bogus but detrimental to a sound moral philosophy. The goal of moral psychology and moral philosophy alike should be understanding and liberation. Moral philosophy will give us the guidance that comes from moral understanding, critical intelligence, and the cultivation of moral imagination. It will not tell us what to do, but it will help us struggle to discern better from worse possibilities within a given situation. Moral philosophy cannot, and never did, give us an adequate theory of moral governance. Once we are liberated from this illusion, we can interpret Kant's dictum—always to think for yourself—as a call for a mature attitude of continual, well-informed, critical, and imaginative moral experimentation. And in our ongoing communal moral experimentation, good cognitive science, coupled with the cultivation of moral imagination, should lead the way.

Notes

1. Alan Donagan, *The Theory of Morality* (Chicago: University of Chicago Press, 1977), pp. 1, 7.
2. Owen Flanagan, *Varieties of Moral Personality* (Cambridge, Mass.: Harvard University Press, 1991), p. 32.
3. Samuel Scheffler, *Human Morality* (Oxford: Oxford University Press, 1992, p. 8).
4. John Dewey, *Human Nature and Conduct* (1922), in Jo Ann Boydston, ed., *The Middle Works of John Dewey, 1899–1924* (Carbondale: Southern Illinois University Press, 1988), pp. 204–205.
5. A large number of these studies of the metaphoric nature of human conceptual systems are discussed in George Lakoff, *Women, Fire, and Dangerous Things: What Categories Reveal about the Mind* (Chicago: University of Chicago Press, 1987), and in Mark Johnson, *The Body in the Mind: The Bodily Basis of Meaning, Imagination, and Reason* (Chicago: University of Chicago Press, 1987).

6. Most of the relevant work on the metaphoric nature of our moral concepts is summarized in Mark Johnson, *Moral Imagination: Implications of Cognitive Science for Ethics* (Chicago: University of Chicago Press, 1993), esp. chap. 2.

7. In ibid., chap. 3, I have shown how such abstract and formal moral principles as the various formulations of Kant's categorical imperative are actually based on different conceptual metaphors, without which there could be no application of moral laws to concrete situations.

8. Ibid.

9. Dewey, *Human Nature*, p. 165.

Chapter 4

Whose Agenda? Ethics versus Cognitive Science

Virginia Held

Ethics and Science

I will argue that cognitive science has rather little to offer ethics, and that
what it has should be subordinate to rather than determinative of the
agenda of moral philosophy. Moral philosophers often make clear at the
outset that moral philosophy should not see the scientific or other explanation of behavior and moral belief, or the prediction and control that science
has aimed at, as our primary concerns. Our primary concern is not explanation but recommendation. I start from this position: ethics is normative
rather than descriptive.

In addition to ethics in its most general form, we need inquiries in all the
more specific areas where ethical considerations should guide us. We need
to be aware of the natural human tendencies and empirical realities that
make moral recommendations feasible or not. We need to specify normative recommendations for particular types of cases, values that will be
suitable for teaching and practicing, and so forth. So we ought to have
inquiries we could call moral sociology, moral psychology, moral economics, moral political science, moral health sciences, and so on (Held 1984).

In these areas, the normative and the descriptive would intermingle. But
it might be unwise to think of the sum of all of these—ethics plus moral
social science, and so on—as "naturalized ethics." One reason is that it
would misrepresent these fields. Another is that in the culture in which we
live and work, the normative is constantly in danger of becoming swallowed by the empirical, not just in the social sciences themselves but even
within philosophy. Areas such as moral economics, moral sociology, and
so on do not even exist as "fields" in the academy, although, of course,
moral assumptions are necessarily being smuggled into the social sciences
unacknowledged much of the time. It is true that many political science
departments may have one political theorist who may be interested in the
moral norms of political theory. But how many economics departments
have a comparable scholar engaged in inquiry about moral theory in the
area of economics? To help us keep our inquiries in perspective, we might

compare the number of academic appointments in ethics with the number in the empirical social sciences and psychology combined. The important task for moral philosophy in relation to these other areas seems to be to keep the distinctively normative alive and well.

Certainly moral philosophy should not ignore cognitive science. Moral philosophy needs to pay attention to findings in psychology, as to those in economics, sociology, anthropology, and so forth. It should, however, put no special premium on psychology. What ethics should not do is to lose sight of the distinctive and primarily normative and evaluative agenda for which it should continue to press.

Many of us are not arguing for a pure, rationalistic conception of normative ethics—the kind of theory disparaged by many of those enthusiastic about the contributions that cognitive science can make to ethics. We argue for a complex but meaningful and important extension of genuinely normative moral philosophy into many domains now restricted by their primarily empirical agenda. Moral philosophy needs to understand the empirical realities in all these areas, but it should set its own goals and recommend that in all these various fields we not only include a normative component along with everything else that is to be explained but that we give priority to our normative aims. The key is whether what is primarily being sought is causal explanation or moral guidance.

The search for causal explanation dominates in cognitive science and in moral psychology as usually pursued. It has led to various attempted takeovers even of moral philosophy itself. If those interested in moral psychology want it to be genuinely normative (as some do), then our positions are not far apart. But most of the interest in the normative shown by those engaged in the kind of moral psychology influenced by cognitive science is an interest in the explanatory role of moral beliefs, not the validity of moral claims.

If moral psychology is the psychology of making moral judgments and developing moral attitudes, it seeks causal explanations of how this is done. This leaves unaddressed the normative questions of whether the positions arrived at are morally justifiable. If, on the other hand, moral psychology deals with how we ought to cultivate the right kinds of moral attitudes and tendencies, to achieve the morally best outcomes, and to express the most morally admirable ways of living human lives, then it is a branch of moral philosophy. It is moral philosophy of a particular kind, making recommendations in a particular region, the region of psychological traits and responses and of learning, as distinct, say, from the regions of organizing economic life or shaping political institutions. But it gives priority to a normative agenda.

This issue is a separate one from the relation of theory to practice. The relation of moral theory to moral practice should not be thought of as deductive, as simply a matter of applying theory to practice. I have long argued that in ethics, practice should inform theory as well as theory inform practice (Held 1984, 1993). This is an issue that can be argued within both ethics and cognitive science. It applies to normative theory and normative practice, on the one hand, and to empirical theory and empirical practice, on the other. In this chapter, I am dealing with a different issue, though the two are often confused.

Adherents of the view that cognitive science can greatly advance moral inquiry usually reject the fact-value distinction as a mistake (Johnson 1993, 184; Flanagan 1984, 245). We can agree that the line between fact and value is neither sharp nor stable but still hold that there are important differences between clear cases of fact and clear cases of value. To know the caloric intake per day that a child needs to survive is different from deciding that we morally ought to provide these calories. Many of us think there are good reasons to assert such distinctions. In discussing the norms of rationality, Allan Gibbard notes that a "descriptivistic analysis" of them cannot be satisfactory: "It misses the chief point of calling something 'rational': the endorsement the term connotes" (Gibbard 1990, 10). This is even clearer for the norms of morality than for those of rationality.

What some of us would like to see satisfactory explanations of is why so much of the dominant philosophy of the twentieth century in the United States and England has been so subservient to science. These explanations would be in terms of the sociology of knowledge and of psychological explanation. They would require an understanding of the politics of inquiry and the politics of education. They would ask why it is that so many philosophers have devoted themselves to fitting whatever they wanted to say about mental states and human experience and moral choice and evaluation into the framework of scientific description and explanation. These inquiries would ask why philosophy has so seldom resisted rather than accepted domination by the intellectual outlook of science. Thomas Nagel writes about how philosophy is currently "infected by a broader tendency of contemporary intellectual life: scientism. Scientism is actually a special form of idealism, for it puts one type of human understanding in charge of the universe and what can be said about it. At its most myopic it assumes that everything there is must be understandable by the employment of scientific theories like those we have developed to date" (Nagel 1986, 9). But, of course, the scientific outlook is just one of a long series of historical outlooks and is, in Nagel's view, "ripe for attack" (Nagel 1986).

Ethical naturalism has lately been presented as a view hospitable to cognitive science, one that avoids such metaphysically peculiar entities as moral norms or normative properties. Ethical naturalism is seen as having what is taken as the great advantage of being consistent with science and metaphysical materialism. But as Jean Hampton usefully reminds us, such a view is hardly new; it has already had an eloquent exponent in Hobbes. She shows how even Hobbes "cannot keep normative standards of value and reasoning out of his theory" (Hampton 1992, 350) and argues that since contemporary naturalists employ objective normative standards for such matters as instrumental rationality and coherent, healthy preferences, they have no defensible grounds for dismissing objective moral norms.

An additional reaction might be that since so many theorists and students of ethics recognize the reasons to move beyond Hobbes's naturalistic reduction of the moral to the empirical, they should be able to see how similar arguments apply to contemporary versions of ethical naturalism. When the more sophisticated machinery of philosophy of mind and philosophy of language is used to construct ethical naturalisms with Hobbesian premises, the results are not substantially less vulnerable than Hobbes's own construction employing these premises.

Moral Experience

Suppose we start with the other fork of the Kantian antinomy between, on the one hand, the freedom we must assume to make sense of moral responsibility, and explanation and prediction in terms of the causal laws of nature. Then we might hold that moral philosophy should make room, first of all, for moral experience and its mental states as these appear to the moral subject rather than to an observer of that subject.

Experience should most certainly not be limited to the empirical observation so much philosophy has seen it as equivalent to. Moral experience includes deliberation and choice and responsibility for action; it includes the adoption of moral attitudes, the making of moral judgments about our own and others' actions and their consequences, and evaluation of the characters and lives we and others have and aspire to. And it includes these as experienced subjectively (Held 1984, 1993).

Moral experience includes deciding what to do. It requires us to assume we can choose between alternatives in ways that should not be expected a priori to be subsumable under scientific explanations, whether psychological, biological, ecological, or any other.[1] I have a different view of moral experience than did Kant, but I share the view that moral experience requires assumptions that cannot be reconciled with causal explanation as so far understood and that this should not lead us to give up the distinction between recommendations that we can accept or reject and explanations of what it is that we do.

Moral experience finds us deliberating about which moral recommendations to make into our reasons for acting, and reflecting on whether, after acting, we consider what we have done to be justifiable. It finds us weighing the arguments for evaluating the actions of others one way or another, and evaluating the states of affairs we and others are in and can bring about. It finds us approving or disapproving of the traits and practices we and others develop and display.

My view of moral epistemology is in some respects rather like that of many ethical naturalists: I agree that the picture of convergence and progress in science as contrasted with continued disagreement and lack of progress in ethics is a distorted one (Flanagan 1994; Churchland 1989). In the area of moral inquiry, there are what can be thought of as the equivalents of observation statements: they are the particular moral judgments whose acceptance allows us to "test" our moral principles, which principles can be taken as comparable to hypotheses. There is thus to some extent a parallel between moral and scientific inquiry. But in another way moral inquiry is fundamentally unlike scientific inquiry: the objective moral norms we can come to think of as valid are not about "natural" entities or events, and whereas with observation statements we are trying to report what the world causes us to observe, with moral judgments we are trying to choose how we will act in and upon the world (Held 1982, 1984, chaps. 4, 15; 1993, chap. 2).

We need a conception of mind that allows us to make sense of moral experience, moral choice, and moral evaluation. Cognitive science and conceptualizations consistent with it are highly influential in philosophy. Huge amounts of philosophy of mind are now devoted to the task of constructing conceptions of mind compatible with cognitive science. But those of us interested in ethics should insist, I think, on pursuing our agenda: finding a conception of mind compatible with what we understand and have good reason to believe about moral experience. If this conception of mind is incompatible with cognitive science, so much the worse for cognitive science. It has only been so much the worse for ethics because we have let the agenda be set by those who have little, or only marginal, interest in ethics. I thus subscribe to Thomas Nagel's recommendation that when the subjective and objective standpoints "cannot be satisfactorily integrated ... the correct course is not to assign victory to either standpoint but to hold the opposition clearly in one's mind without suppressing either element" (Nagel 1986, 6).

The program of the kind of philosophy of mind that has grown out of or is compatible with cognitive science has been to study the causal role of mental phenomena. Among the mental phenomena to be studied may be moral beliefs, intentions, and preferences. But the aim is to see their causal roles in the explanation of other mental phenomena and of behavior. In Alvin Goldman's useful summary, functionalism, like behaviorism, "tries to account for the meanings of mental expressions ultimately in terms of external events (stimuli and behavior) without succumbing to behaviorism's pitfalls. Unlike behaviorism, functionalism acknowledges that mental states are 'inner' states. But it is not the *intrinsic* quality of an inner state that makes a mental state what it is. Rather, it is the state's *causal relations*: first, its relations to external events such as stimulus inputs and behavioral outputs and second, its relations to other internal states" (Goldman 1993b, 72).

This kind of philosophy of mind speaks of "mental" and "inner" states, but they are not mental in the sense of subjective. David Lewis, for instance, describes the internal state a combination lock is in when it is in a state of "unlockedness," as if a mind's internal state could be discussed by analogy with this, although a lock clearly does not have a subjective perspective the way a mind does (Lewis 1966, 17–25). Functional accounts of the mind typically include descriptions of how what are called mental states are related to other mental states, but, as Goldman notes, the internal state concepts "are jointly explained in terms of their (causal) relations to external events and to one another" (Goldman 1993b, 73).

Many others in philosophy of mind go even further to eliminate the subjective, holding that neural events and neural properties are all there is to the mind. According to recent reports, cognitive scientists have made great strides in understanding the region of the brain used for moral beliefs and deliberation. A recent finding along these lines was hailed by Patricia Churchland as "tremendously important" for what we think about "moral character, empathy and the determinants in choosing right over wrong" (*New York Times* 1994, C1). But why should we think such findings of importance to ethics? We already know a great deal about the physical location of our ability to see, about how the eye works and how vision takes place. Yet all this knowledge tells us nothing about which patterns of colors have aesthetic merit and which do not. We already know that to have moral beliefs requires a brain, and that if the brain is impaired, this may lessen the degree to which we think we should hold persons responsible. But what we think persons ought to do when they are responsible is a normative question that no amount of further knowledge of how the brain works can address. And just as knowing how the eye works and explaining how vision occurs do not get at the felt, conscious awareness of seeing a red apple or at understanding the validity of an aesthetic judgment, neither would knowing where in the brain moral deliberation takes place and explaining the causal mechanisms involved be able to get at the felt conscious experience of engaging in moral deliberation, nor at understanding the validity of a normative recommendation.

Thomas Nagel concludes that "the attempt to give a complete account of the world in objective terms detached from [subjective] perspectives inevitably leads to false reductions or to outright denial that certain patently real phenomena exist at all" (Nagel 1986, 7). He sees "this form of objective blindness" as "most conspicuous in the philosophy of mind, where one or another external theory of the mental, from physicalism to functionalism, is widely held." It is clear, he thinks, that "the subjectivity of consciousness is an irreducible feature of reality—without which we couldn't do physics or anything else," including cognitive science, "and it must occupy as fundamental a place in any credible world view" as the entities recognized by the sciences (Nagel 1986, 7–8). A view such as this must be presupposed, I believe, by ethics.

When Hilary Putnam predicts that cognitive science will go the way of Comte's dream of social physics, we may agree for different reasons from those of many who share his view. Putnam thinks the program of functionalism is utopian and inherently incapable of empirical verification (Putnam 1993). We think we should never have wanted to reduce human experience to what can be explained scientifically. We should, on the contrary, assert with conviction that in ethics we seek recommendations rather than explanations. And in philosophy of mind we should seek a conception of mind compatible with subjective moral experience, and with expression and objective moral norms, rather than a conception of mind compatible with scientific objectivity and causal explanation. If our conception of mind in ethics clashes with our conception of mind in science, we can decide to live with that incompatibility rather than abandon our understanding of moral experience and human expression. The domains of literature and art, in facilitating imaginative identification with others, and in providing vicarious moral experience, are at least as relevant to ethics as is science, including cognitive science. If ethics must choose between the perspectives of science and art, it should not reject art.

Philosophers influenced by cognitive science seem to say we do not need to choose: we can use art and literature and anything else to provide data for moral beliefs. But they interpret art and literature as providing material to be studied by science, whereas an aim of art may be to express a unique perspective or to escape scientific explanation. And art must often reflect a subjective point of view not open to any observer. I agree we should not choose between art and science, but that requires us to live with incompatible views, not to reduce art—or ethics—to what can be studied by science and "accounted for" by causal explanation.

Evaluation as Explanation

We have an example of interpreting evaluation as explanation in the recent work of such philosophers as Peter Railton, Richard Boyd, and Nicholas Sturgeon, who call themselves "moral realists" (Railton 1986; Boyd 1988; Sturgeon 1984). And we can see the limitations of this approach for answering the normative questions of moral philosophy.

Instead of calling this recent form of ethical naturalism "moral realism," David Copp advocates the term "confirmation theory," since there are other forms of moral realism with quite different positions (Copp 1990). I find the suggestion appropriate and useful. Confirmation theory, then, holds that moral theory can be empirically confirmed.

Copp argues persuasively that we can always doubt "that the explanatory utility of a set of postulates has any tendency to justify, certify, or ground a moral standard" (Copp 1990, 239). Moral codes are not explanatory theories, and hence confirmation theory, and the type of moral realism it represents, are misguided. Copp's well-formulated arguments apply as well to the view that cognitive science can confirm, or give us reason to reject, moral theories or recommendations, so let us consider some of them.

Confirmation theorists hold that certain moral facts provide the best explanation for certain empirical findings. Nicholas Sturgeon, for instance, suggests that the moral fact that Hitler was evil explains his behavior in ordering the "final solution" (Sturgeon 1984). The confirmation theorist, Copp notes, "wants to say that our moral beliefs are justified because we need them for explanations" (Copp 1990, 242), just as in the empirical sciences, the explanatory utility of an empirical theory is evidence of its truth. But, Copp argues, we can "explain our having the considered moral beliefs we have without assuming their truth, by citing [for instance] the facts of acculturation and other relevant sociological and psychological facts" (Copp 1990, 242).

Consider another of Copp's examples. Someone might seek to "explain Stalin's ruthless behavior on the basis that he was an (approximation to the) overman" of the Nietzschean moral perspective (Copp 1990, 247). But such an explanation would be a purported psychological one; we "would not be tempted in the least to accept Nietzschean morality on this basis" (248). We would not suppose that the explanatory utility of the concept of the Übermensch would justify taking it as a moral standard.

Let me add a further argument. Consider Peter Railton's claim that the presence of social justice (a normative judgment) can help explain an absence of social unrest (an empirical finding), and an absence of social justice can help explain the occurrence of social unrest (Railton 1986). We can all agree that an absence of popular dissatisfaction can contribute to social cohesion and an absence of unrest, but we need not agree that any moral judgment is confirmed by such an explanatory claim. An absence of unrest might be brought about by the violent suppression of dissent. Suppose we have two societies in which an equal amount of cohesion and absence of unrest obtain for an equally long period. In one society, this is caused by arrangements that satisfy the needs of its citizens, in the other by ruthless suppression of dissent. And suppose that these are the only significant causal factors and results that differ in the two cases. Can we conclude anything about the moral value of social justice from these causal explanations? No. But does the failure of either factor to provide a better explanation than the other mean we cannot make a moral judgment about which is the morally better way to prevent unrest? Again, no.

Paul Churchland claims that we can explain the learning of moral facts the way we can explain the learning of scientific facts. In both cases what counts is practical wisdom, learning how. And in both cases, the body of learning can be improved, refined, made more unified. Churchland speaks of the replacement of one prototype by another. The prototype of the stern father as moral authority, where others are seen as siblings, is replaced, for instance, by the "more arresting account" of persons as parties to a contract (Churchland 1989, 303). But this is not a process that happens as if by itself. We decide what prototype to make "more arresting," that is, which to consider morally or descriptively better. We can try to explain why such transitions have occurred. One common explanation for the one Churchland considers has to do with the replacement of feudalism by capitalism. But such explanations do not tell us whether the replacement of one moral paradigm by another or one socioeconomic system by another was or would be morally justified. Only an ethics that is not a branch of cognitive science can even aspire to do so.

Some may think, with Geoffrey Sayre-McCord, that even transconventional rules, like those of logic, rationality, or morality, can provide normative explanations of behavior. On this view it could be shown how a feedback mechanism "by and large and in standard conditions rewards compliance and penalizes non-compliance" (Sayre-McCord 1988, 67). Although we cannot yet know which moral theory offers the best explanation of people's moral beliefs and behavior, in principle we may be able to find out. And suppose we could? As Sayre-McCord himself suspects, we cannot decide whether a moral theory is justifiable on the basis of how well it explains belief and behavior. And, we may add, to decide that what people perceive as rewards and penalties are the appropriate grounds on which to consider a moral theory justifiable would itself be a matter of moral theory. Do people's valuing lives of indolence or, alternatively, lives of ascetic denial make either kind of life morally admirable?

Perhaps it is possible to explain the popularity of confirmation theory. In various contemporary philosophical circles, only those claims consistent with what can be confirmed scientifically are deemed worthy of philosophical attention. But that this approach is so unsatisfactory for dealing with the whole domain of morality should lead to its rejection.

Some philosophers think that a naturalistic view will avoid certain characteristic distortions with which many other moral theories have presented moral life. Annette Baier, for instance, discusses male concepts of the person that rest on denials that a person *qua* person need be born at all or need be a family member. Moral theories built on agreements that might be reached by purely rational beings, or on notions of rationally maximizing individual preferences, often assume such conceptions of the person. Baier observes that it is "unlikely that women, who have traditionally been allocated the care of the very dependent young and old persons, will take persons as anything except interdependent persons" (Baier 1990, 11). Women are not likely to forget that persons cannot come into existence any other way than by being born of women. But this issue, in my view, is independent of whether ethics should be naturalized.

Feminist reconceptualizations of the person are having a profound influence on feminist moral philosophy. The persons whom morality is for will need quite different moralities from the currently still dominant ones, depending on what persons are understood to be. And these will be moralities for both supposedly public persons and supposedly private persons, since they are, after all, the same persons.

This large and deep revision of moral concepts and theories has not needed much assistance from cognitive science. The latter can sometimes strengthen one suggestion or another, as when it offers evidence that self-interest is in fact not the only motive determining human behavior. But evidence for this has always been all around us in our moral experience, if we would be willing to see it. And women will often do well to trust our own experience when it conflicts with the assertions of science, cognitive or other. The history of psychological and social scientific misinterpretation of women's experience is a horror story not to be forgotten.

Empathy and Ethics

Alvin Goldman has provided a useful summary of work in cognitive science on various aspects of empathy and on the role of empathy in influencing moral behavior. He notes that a satisfactory prescriptive theory should be "rooted in human nature" and thus that findings in cognitive science are relevant to normative moral theory (Goldman 1993a, 357). We can all agree that a normative moral theory should not ask the impossible of human beings and that findings in cognitive science can inform us of emotional tendencies and limitations that may be relevant to what we ask morality to recommend. But we should not exaggerate the contributions of cognitive science.

We can study the tendencies of human beings, especially some of them, toward aggression and violence. Ethics needs to recommend how to counter and limit such tendencies. We can study how stereotypes are formed and how prejudices develop; ethics should instruct us on what fairness and consideration require in opposition to such prejudices. Knowing what tendencies people have and how attitudes develop is useful knowledge surely, but it does not do the work of moral evaluation and recommendation.

A neural network model, for instance, may explain how children come to learn various moral attitudes and to make interconnected moral judgments (Churchland 1989). But it may be as useful for explaining how children learn to disparage weaker groups and to take advantage of their members as it is for explaining how admirable moral attitudes are learned. It does rather little to answer questions about what we ought to teach our children, though once we have decided what we ought to teach them, it may help us understand how to do so more effectively. Paul Churchland writes that on the basis of the neural network models of cognitive function now emerging, "knowledge acquisition is primarily a process of learning *how*: how to recognize a wide variety of complex situations and how to respond to them appropriately ... moral knowledge does not automatically suffer by contrast with other forms of knowledge" (Churchland 1989, 298). But questions about which responses are appropriate and which are not are just those ethics must address, and before we can decide what moral "knowledge acquisition" should be promoted and what beliefs and attitudes should be discouraged, we need to evaluate alternative moral norms.

The facts of how human beings can and do feel empathy are certainly relevant to morality (Meyers 1993). But many of these facts have been as plain to those wishing to see them as the facts of how human beings can and do reason. Evidence of empathy and its role in benevolent action has been ubiquitous, but it has often not been noticed by moral philosophers or it has not been considered of importance for morality. Some of us find it extraordinary that only after cognitive science informs us that empathy exists, and can motivate people to act in ways that are not primarily a matter of satisfying their own individual interests, is this believed by many philosophers. We may wish to ask: what can be the explanation of this?

One factor helping to explain why the facts about empathy have not been noticed, or have not been deemed relevant to morality, has been the deep association between reason and men and between emotion and women, and the discounting of the experience of women. The moral experience of women offers a vast source of insights into the phenomena of empathy and how it may be excessive at times, as well as deficient in other cases. These insights have already greatly affected feminist moral theory, with no particular need to make use of cognitive science.

One can point to a long history of denying the evidence when the evidence was presented by the actions, motives, or experience of women. The whole realm of women's actions and experience within the family and in the household has been dismissed as not relevant to morality, as not even moral action but rather as behavior belonging to the realm of the biological. Women have been conceptualized as merely carrying out the dictates of their maternal natures rather than exercising moral judgment.[2] Even those who hold that all action is determined but that we can think of some of it as relatively free if self-determined have not conceptualized women's mothering activities as offering examples of action that might be free in this sense. Rather, they too have usually seen it as instinctive behavior, not action guided by and relevant to the revising of moral principles. Feminist theorists point out the distortions in nearly all traditional and standard moral theories. Such theories give, in the construction of morality, unjustified privilege to the public domain where men have been predominant, and unreasonable disadvantage to the private domain to which women have largely been confined (Held 1993).

Cognitive science can confirm that action within the household, like much other action, is often determined by natural emotions. But cognitive science cannot make the case that there are moral choices to be made about what sorts of actions within the family and between friends are to be recommended as distinct from observed and explained, and that experience in this domain is highly relevant to moral theory. At the moment, it seems to be only feminist moral theorizing that is taking this moral experience of women seriously. And the aims of feminist moral theory and of cognitive science are very different and sometimes at odds.

Goldman usefully points out that empathy need not be merely particularistic, extending to those we are close to or who are like ourselves. Although there are certainly limits to our benevolence, we can feel empathy for persons quite distant from us, and there are "sympathy-based theories that are quite universalistic" (Goldman 1993, 358). But cognitive science can do little to guide us in deciding how universalistic our moral theories ought to be. Owen Flanagan is content to have them more local (Flanagan 1991). Others find moral theories deficient unless they are universal. Cognitive science can suggest limits on what we deem reasonable to expect of persons, and it can certainly help us to teach and to gain compliance with the moral norms we deem best, but it can not itself assist us with decisions about which moral norms to adopt and which moral theory is best. Within feminist moral theorizing, there is much disagreement, ample for healthy debate, including disagreement about how universalistic and how naturalistic ethics should be. My own view is, of course, not *the* view of feminist moral theory. But no moral theory can be acceptable from a feminist point of view unless it includes a strong moral commitment to gender equality and an end to gender dominance and unless it pays adequate attention to the experience of women. Some of these commitments, in my view, escape naturalistic description.

Metaphor and Philosophical Thought

Many of us agree with Mark Johnson that our basic moral concepts are importantly metaphoric (Johnson 1993). But understanding this does not lessen our need to make moral choices between alternative metaphors. Consider some examples. For centuries, the concept of "human" has been based on images of a male human being, often a warrior, or on images reflecting other aspects of what men do when they exercise political power or create works of art or science or industry. Feminist critiques of prevailing conceptions of the human as "man" insist that the concept include images of "woman" and of "girl," and in reasonable ways rather than always as an "other".

We can see in the history of ethics and social and political philosophy overwhelming reliance on the metaphor of man in the public sphere contracting with other men to limit the means by which male heads of household pursue their own interests, in political life and in the marketplace. Once we become conscious of such dominant metaphors, through perceptive interpretations of them, we can decide to expand them or modify them or limit them or replace them. This has been a continual task of philosophy.

I have discussed in various places the concept of property and how Lockean images of the independent farmer or tradesman who can acquire what he needs if only he is not interfered with still permeate contemporary interpretations of property. These metaphors are woefully inadequate for the task of evaluating economic arrangements, especially in contemporary economic systems (Held 1984). I have also discussed how the image of the orator in the square who can speak if only not prevented still guides interpretations of freedom of expression and First Amendment decisions in the law, though this metaphor is grievously unsuited for dealing with contemporary cultures, such as our own, that are dominated by the realities of the media and where getting a hearing through the media bears very little resemblance to speaking out in the public square (Held 1988). And I have argued that the metaphors of economic exchange are not the right ones for dealing with many moral issues where they have been employed routinely (Held 1984; see also Anderson 1993). I am clearly highly sympathetic with efforts to clarify the metaphors used in moral thinking. However, once we have understood how deficient reigning metaphors often are, we need to choose alternatives that will be morally and descriptively better.

Some philosophers hold that the metaphor of the contract and its ways of thinking should be applied not only to a social contract underlying political institutions but to most or all human relationships and even to the whole of morality (Gauthier 1986; Hampton 1993). Others argue, on the contrary, that contractual thinking is very unsuitable for thinking about large parts of morality and about much that matters in families and societies. The arguments between us on these issues cannot be settled by examining how people learn, how they come to think what they do, or how contractual norms and ways of thinking explain what people believe or do. The argument some of us are making is that contractual thinking is rampant in the society around us but that it should not be as pervasive as it is. In our view, this metaphor is not appropriate for a great deal of moral understanding. Cognitive science cannot recommend such choices among metaphors, though it may help us explain why the choices that have been made have been accepted, and other empirical studies will help explain why the choices made have so often been advantageous to some and disadvantageous to others.

Mark Johnson thinks we are just beginning to understand, by means of cognitive science, how the metaphors we use influence the ways we think about morality. This seems to me wildly to overstate the contributions of cognitive science. Philosophers and others frequently illuminate and examine the dominant metaphors of various outlooks and ways of thinking. One can think of such examples as C. B. Macpherson's "possessive individualism" in Hobbes and Locke, the use of Isaiah Berlin's hedgehog and fox images to distinguish the outlooks of continental and British philosophy, Nancy Hartsock's analysis of the metaphor of the barracks community in thinking about politics, and countless others (Macpherson 1972; Berlin 1953; Hartsock 1983). Carolyn Merchant suggests some implications of the metaphors by which the earth was conceptualized and thought and felt about. Before the sixteenth century, the earth had been seen as a generative and nurturing mother; there had been a reluctance to violate it, and commercial mining was thought morally problematic because of the way it cut into the earth. By the seventeenth century, sexual metaphors had become prevalent. The earth was thought of as a woman to be dominated by the new science, a female who should appropriately be mastered. Penetrating the earth through mining was no longer seen as a moral problem (Merchant 1982).

When persuasive, such interpretations of prevalent metaphors surely heighten our understanding; they need no support from cognitive science. We are all familiar with how illuminating it is to see the mechanical model at the heart of a great deal of philosophical and other thought and then to question its appropriateness. Or we recognize the metaphors of evolution and survival of the fittest at work, often for dubious purposes, in other bodies of thinking, and we have done this unaided by the lenses of cognitive science. Now, of course, we can see the metaphor of the computer and its programs, its hardware and its software, as pervasive in the philosophical and other thought affected by cognitive science. Some think that to use the metaphor of the computer and its programs for understanding moral experience is surely less than helpful. Thomas Nagel writes that "eventually ... current attempts to understand the mind by analogy with man-made computers ... will be recognized as a gigantic waste of time" (Nagel 1986, 16).

Mark Johnson seems to think that the logic of the metaphor used determines our reasoning in what is called the "target" area, in this case, the area of ethics. But though it certainly colors it, we have to decide whether any given metaphor leads to good or poor reasoning in a given area, or to a better or worse interpretation of experience in it. When we decide the metaphor is unsuitable, we search for an alternative. Even if a metaphor gives rise to a certain way of thinking, this does not mean that the way of thinking in question can be evaluated on that basis. The distinction between how we happen to come to a position and what makes the position true or false or valid or invalid or a good or bad way of thinking about an issue is, of course, a distinction with a very long history. Many of us think there are good reasons to maintain it.

Cognitive science may suggest that in answering questions about how we come to think that we should keep our agreements, we answer questions about whether we should, and the same for the other moral judgments it looks at. But we can certainly disagree. It is the task of ethics to answer questions about what we ought to do and what has value. Knowledge about how we came to have the views we do can certainly be helpful, but it can never itself answer moral questions. We have to decide which metaphors are best for dealing with our values, our goals, our obligations. A quite different way of interpreting what is going on from the way Mark Johnson interprets it is to think that in ethics we start with various moral intuitions, judgments, and principles and use metaphors to try to express them.

Let me offer what will be something of a parody of the approach of cognitive science. Mark Johnson singles out various basic metaphors from what he sees as a physical domain and shows how, in his investigations, various types of moral discourse seem to be determined by them. Without the benefit of cognitive science, let me suggest two other basic metaphors from the ones he finds that can easily be seen in moral thinking and discussion.

The first is the SEXUAL/HUNGER metaphor. We have wants and desires, whose satisfaction is pleasurable; it is good to achieve the satisfaction of desire. What we ought to do is to maximize satisfaction. When we are sated, the pleasure decreases; the pleasure is most intense when our desire is greatest. Sometimes we desire the desire even more than the satisfaction.

Pleasure is desired and desirable; pain is avoided and undesirable. It is good to minimize frustration, though we should not be misled by our short-term wants. We should consider what goals to pursue and that there are second-order desires as well as immediate ones.

To think of the SEXUAL/HUNGER metaphor that can be discerned in such discourse as coming from a basic physical domain and that this metaphor determines what can be said in the moral domain employing it may well be misleading. We often start with a mix of moral thoughts and feelings, attitudes and judgments. We try to express them using metaphoric language. It might be more accurate, at least sometimes, to say that the moral terms belong to a basic domain and the physical ones, used in the context of moral discourse, to what could be called, in this parody, a "vehicle" domain.

According to a second metaphor, the GROWTH metaphor, the normal person develops moral understanding. When moral growth is interfered with, the person may be morally stunted. Without an environment that facilitates moral growth the person may be morally deformed. Noxious conditions sometimes impede natural moral development. Morality enables persons to grow as human beings.

One could say of this way of speaking that in the basic, moral domain, growth is given a moral interpretation; in the "vehicle" domain it is given a biological one. Philosophers who have relied on the GROWTH metaphor have assumed rather than argued that growth is good. Now that we find it easy to question this in areas such as population growth and the effects of industrialization on the environment, it may be easier for us to see what is wrong with such a metaphor. But one could see it, as many did, long before the questioning of growth as good became popular.

One could go on, designating other "vehicle" domains and drawing word pictures of them. Of course, metaphors can be discerned in moral discourse without the aid of cognitive science and without suggesting we need a research program in "vehicle domains." Moral deliberation is an imaginative enterprise, we can agree. But there may be more in literature and art and in listening to others with experiences like and unlike our own to nourish this imaginative enterprise than in cognitive science, unless, to conclude our story, the latter becomes so fanciful a narrative that it is no longer recognizable as science.

Moral Philosophy

John Dewey's ethics remain fundamentally unsatisfactory, in my view, because he thought moral theory was the sort of theory to which sciences like cognitive science could provide answers.[3] He wrote as if moral problems simply present themselves and as if the tasks of morality are to find empirical solutions to such problems. Moral problems, however, do not simply present themselves. We decide to make certain empirical situations into moral problems, to interpret them as moral problems. Forty years ago, very few people saw the prevalent confinement of women to household roles as a moral problem, a matter of injustice. On the contrary, dissatisfaction with such confinement was interpreted as a personal problem of psychological adjustment. But because a few persons, later joined by others, made the normative as distinct from empirical judgment that it was unjust that opportunities outside the household were so widely closed to women, opportunities have gradually increased.

Many of us are not arguing for a morality of rules, either Kantian or utilitarian. These rule moralities can be interpreted as generalizations to the level of moral theory of the norms of law and public policy. Many feminists have criticized the excessively rationalistic moral theories that have been dominant in recent decades. We argue that we should seek a morality that adequately considers the activities, emotions, and values involved in caring and that takes these as seriously as moral theories have taken justice and reason and utility. But we may still think that moral philosophy should give priority to the normative over the empirical, to recommendation and evaluation over explanation and prediction.

Of course, if one supposes that the categories into which everything conceivable must fall are the "natural" and the "supernatural," it is easy to think that ethics belongs in the realm of the natural. I would think this also if I accepted this division, but to take it as presented seems highly misleading. The division between the natural and the supernatural may be one of the most unfortunate metaphors of all. There are other distinctions that may be far more useful, especially for ethics: the distinction between that which is specifically human and that which belongs to a natural world that would be much as it is with no humans in it, or the distinction between the subjective point of view of the conscious self and the objective point of view of the observer studying nature and the human beings in it and seeking explanations of its events. These distinctions require us to recognize realms that are not supernatural but are not natural either in the sense just suggested.

It is to these other domains of conscious human subject, constantly shaping and being shaped by the social relationships that are part of what he or she is, that normative recommendations are addressed. We do not address our moral recommendations to the nature studied by the natural sciences but to persons consciously choosing how to live our lives, change our surroundings, express our hopes, and continue or perhaps even improve the futures of human beings in nature.

Notes

1. For a defense of incompatibilism, see van Inwagen (1983).
2. For further discussion and many references, see Held (1993).
3. Dewey recommends, for instance, finding "the facts of man continuous with those of the rest of nature" (Dewey 1957, 13).

References

Anderson, Elizabeth. 1993. *Value in Ethics and Economics.* Cambridge, Mass.: Harvard University Press.

Baier, Annette. 1991. "A Naturalist View of Persons." *Proceedings and Addresses of the American Philosophical Association* 65, no. 3 (November).

Berlin, Isaiah. 1953. *The Hedgehog and the Fox.* New York: Simon and Schuster.

Boyd, Richard. 1988. "How to Be a Moral Realist." In *Essays on Moral Realism.* Edited by Geoffrey Sayre-McCord. Ithaca, N.Y.: Cornell University Press.

Churchland, Paul. 1989. *A Neurocomputational Perspective: The Nature of Mind and Structure of Science.* Cambridge, Mass.: MIT Press.

Copp, David. 1990. "Explanation and Justification in Ethics." *Ethics* 100 (January): 237–258.

Dewey, John. 1957. 1922. *Human Nature and Conduct.* New York: Random House.

Flanagan, Jr., Owen J. 1984. *The Science of the Mind.* Cambridge, Mass.: MIT Press.

Flanagan, Jr., Owen J. 1991. *Varieties of Moral Personality: Ethics and Psychological Realism.* Cambridge, Mass.: Harvard University Press.

Flanagan, Jr., Owen J. 1994. "Ethics Naturalized: Ethics as Human Ecology." Paper presented at Mind and Morals Conference, Washington University, St. Louis, April 8.

Gauthier, David. 1986. *Morals By Agreement.* Oxford: Clarendon Press.

Gibbard, Allan. 1990. *Wise Choices, Apt Feelings: A Theory of Normative Judgment.* Cambridge, Mass.: Harvard University Press.

Goldman, Alvin I. 1993a. "Ethics and Cognitive Science." *Ethics* 103 (January): 337–360.

Goldman, Alvin I. 1993b. *Philosophical Applications of Cognitive Science.* Boulder, Colo.: Westview Press.

Hampton, Jean. 1992. "Hobbes and Ethical Naturalism." In *Philosophical Perspectives,* vol. 6, *Ethics.* Ed. James E. Tomberlin. Atascadero, Calif.: Ridgeview, 1992.

Hampton, Jean. 1993. "Feminist Contractarianism." In *A Mind of One's Own: Feminist Essays on Reason and Objectivity.* Boulder, Colo.: Westview Press.

Hartsock, Nancy. 1983. "The Barracks Community in Western Political Thought." In *Women and Men's Wars.* Edited by Judith Stiehm. New York: Pergamon.

Held, Virginia. 1982. "The Political 'Testing' of Moral Theories." *Midwest Studies in Philosophy* 7 (Spring): 343–363.

Held, Virginia. 1984. *Rights and Goods: Justifying Social Action.* New York: Free Press.

Held, Virginia. 1988. "Access, Enablement, and the First Amendment." In *Philosophical Foundations of the Constitution.* Edited Diana T. Meyers and Kenneth Kipnis. Boulder, Colo.: Westview Press.

Held, Virginia. 1993. *Feminist Morality: Transforming Culture, Society, and Politics.* Chicago: University of Chicago Press.

Johnson, Mark. 1993. *Moral Imagination: Implications of Cognitive Science for Ethics.* Chicago: University of Chicago Press.

Lewis, David. 1966. "An Argument for the Identity Theory." *Journal of Philosophy* 63:17–25.

Macpherson, C. B. 1972. *The Political Theory of Possessive Individualism.* New York: Oxford University Press.

Merchant, Caroline. 1982. *The Death of Nature: Women, Ecology, and the Scientific Revolution.* New York: Harper & Row.

Meyers, Diana Tietjens, 1993. "Moral Reflection: Beyond Impartial Reason," *Hypatia* 8: 21–47.

Nagel, Thomas. 1986. *The View from Nowhere.* New York: Oxford University Press.

New York Times. 1994. May 24, p. C1.

Putnam, Hilary. 1993. At Symposium: *Functionalism.* American Philosophical Association, Eastern Division Meeting, December 28.

Railton, Peter. 1986. "Moral Realism." *Philosophical Review* 95:163–207.

Sayre-McCord, Geoffrey. 1988. "Normative Explanations." In *Philosophical Perspectives*, vol. 6: *Ethics.* Ed. James E. Tomberlin. Atascadero, Calif.: Ridgeview.

Sturgeon, Nicholas. 1984. "Moral Explanations." In *Morality, Reason and Truth: Essays on the Foundations of Ethics.* Edited by David Copp and David Zimmerman. Totowa, N.J.: Rowman and Allanheld.

van Inwagen, Peter. 1983. *An Essay on Free Will.* Oxford: Clarendon Press.

Part II
Moral Judgments, Representations, and Prototypes

Chapter 5
The Neural Representation of the Social World
Paul M. Churchland

Social Space

A crab lives in a submarine space of rocks and open sand and hidden recesses. A ground squirrel lives in a space of bolt holes and branching tunnels and leaf-lined bedrooms. A human occupies a physical space of comparable complexity, but in our case it is overwhelmingly obvious that we live also in an intricate space of obligations, duties, entitlements, prohibitions, appointments, debts, affections, insults, allies, contracts, enemies, infatuations, compromises, mutual love, legitimate expectations, and collective ideals. Learning the structure of this social space, learning to recognize the current position of oneself and others within it, and learning to navigate one's way through that space without personal or social destruction is at least as important to any human as learning the counterpart skills for purely physical space.

This is not to slight the squirrels and crabs, or the bees and ants and termites either, come to think of it. The social dimensions of their cognitive lives, if simpler than ours, are still intricate and no doubt of comparable importance to them. What is important, at all levels of the phylogenetic scale, is that each creature lives in a world not just of physical objects but of other creatures as well, creatures that can perceive and plan and act, both for and against one's interests. Those other creatures therefore bear systematic attention. Even nonsocial animals must learn to perceive, and to respond to, the threat of predators or the opportunity for prey. Social animals must learn, in addition, the interactive culture that structures their collective life. This means that their nervous systems must learn to represent the many dimensions of the local social space, a space that embeds them as surely and as relevantly as does the local physical space. They must learn a hierarchy of categories for social agents, events, positions, configurations, and processes. They must learn to recognize instances of those many categories through the veil of degraded inputs, chronic ambiguity, and the occasional deliberate deception. Above all, they must learn to generate appropriate behavioral outputs in that social space, just as surely as they must learn to locomote, grasp food, and find shelter.

In confronting these additional necessities, a social creature must use the same sorts of neuronal resources and coding strategies that it uses for its representation of the sheerly physical world. The job may be special, but the tools available are the same. The creature must configure the many millions of synaptic connection strengths within its brain so as to represent the structure of the social reality in which it lives. Further, it must learn to generate sequences of neuronal activation-patterns that will produce socially acceptable or socially advantageous behavioral outputs. As we will see in what follows, social and moral reality is also the province of the physical brain. Social and moral cognition, social and moral behavior, are no less activities of the brain than is any other kind of cognition or behavior. We need to confront this fact, squarely and forthrightly, if we are ever to understand our own moral natures. We need to confront it if we are ever to deal both effectively and humanely with our too-frequent social pathologies. And we need to confront it if we are ever to realize our full social and moral potential.

Inevitably these sentiments will evoke discomfort in some readers, as if, by being located in the purely physical brain, social and moral knowledge were about to be devalued in some way. Let me say, most emphatically, that devaluation is not my purpose. As I see it, social and moral comprehension has just as much right to the term "knowledge" as does scientific or theoretical comprehension—no more right, but no less. In the case of gregarious creatures such as humans, social and moral understanding is as hard won, is as robustly empirical and objective, and is as vital to our well-being as is any piece of scientific knowledge. It also shows progress over time, both within an individual's lifetime and over the course of many centuries. It adjusts itself steadily to the pressures of cruel experience, and it is drawn ever forward by the hope of a surer peace, a more fruitful commerce, and a deeper enlightenment.

Beyond these brief remarks, the philosophical defense of moral realism must find another occasion. With the patient reader fairly forewarned, let us put this issue aside for now and approach the focal issue of how social and moral knowledge, whatever its metaphysical status, might actually be *embodied* in the brains of living biological creatures.

It cannot be too difficult. Ants and bees live intricate social lives, but their neural resources are minuscule—for an ant, 10^4 neurons, tops. However tiny those resources may be, evidently they are adequate. A worker ant's neural network learns to recognize a wide variety of socially relevant things: pheromonal trail markings to be pursued or avoided; a vocabulary of antennae exchanges to steer one another's behavior; the occasions for general defense, or attack, or fission of the colony; fertile pasture for the nest's aphid herd; the complex needs of the queen and her developing eggs; and so forth.

Presumably the challenge of social cognition and social behavior is not fundamentally different from that of physical cognition and behavior. The social features or processes to be discriminated may be subtle and complex, but as recent research with artificial neural networks illustrates, a high-dimensional vectorial representation—that is, a complex *pattern* of activation levels across a large population of neurons—can successfully capture all of them. To see how this might be so, let us start with a simple case: the principal emotional states as they are displayed in human faces.

EMPATH: A Network for Recognizing Human Emotions

Neural net researchers have recently succeeded in modeling some elementary examples of social perception. I here draw on the work of Garrison Cottrell and Janet Metcalfe at the University of California, San Diego. Their three-stage artificial network is schematically portrayed in figure 5.1.

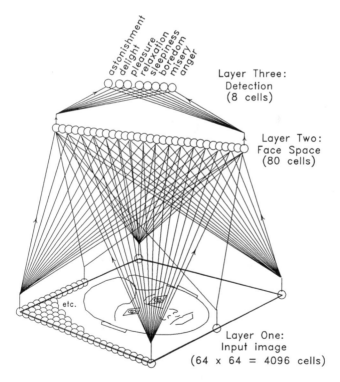

Figure 5.1
EMPATH, a feedforward network for recognizing eight salient human emotions.

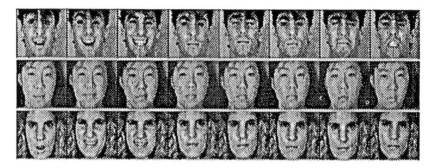

Figure 5.2
Eight familiar emotional states, as feigned in the facial expressions of three human subjects. From the left, they are astonishment, delight, pleasure, relaxation, sleepiness, boredom, misery, and anger. These photos, and those for seventeen other human subjects, were used to train EMPATH, a network for discriminating emotions as they are displayed in human faces.

Its input layer or "retina" is a 64 × 64-pixel grid whose elements each admit of 256 different levels of activation or "brightness." This resolution, both in space and in brightness, is adequate to code recognizable representations of real faces.

Each input cell projects an axonal end-branch to every cell at the second layer of 80 cells. That layer represents an abstract space of 80 dimensions in which the input faces are explicitly coded. This second layer projects finally to an output layer of only 8 cells. These output cells have the job of explicitly representing the specific emotional expression present in the current input photograph. In all, the network contains (64 × 64) + 80 + 8 = 4,184 cells, and a grand total of 328,320 synaptic connections.

Cottrell and Metcalfe trained this network on eight familiar emotional states, as they were willingly feigned in the cooperating faces of twenty undergraduate subjects, ten male and ten female. Three of these charming subjects are displayed eight times in figure 5.2, one for each of the eight emotions. In sequence, you will there see astonishment, delight, pleasure, relaxation, sleepiness, boredom, misery, and anger. The aim was to discover if a network of the modest size at issue could learn to discriminate features at this level of subtlety, across a real diversity of human faces.

The answer is yes, but it must be qualified. On the training set of (8 emotions × 20 faces =) 160 photos in all, the network reached—after 1000 presentations of the entire training set, with incremental synaptic adjustments after each presentation—high levels of accuracy on the four positive emotions (about 80 percent), but very poor levels on the negative emotions, with the sole exception of anger, which was correctly identified 85 percent of the time.

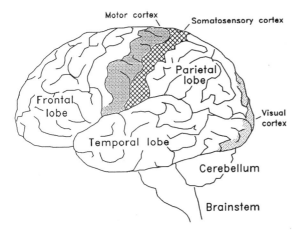

Figure 5.4
The location of some of the primary and secondary sensory areas within the primate cerebral cortex. (Subcortical structures are not shown.) The "motor strip" or motor output cortex is also shown. Notice the broad cortical areas that lie outside these easily identified areas.

simply, recall your teenage years. Mastering that complexity is a cognitive achievement at least equal to earning a degree in physics. And yet with few exceptions, all of us do it.

Are There "Social Areas" in the Brain?

Experimental neuroscience in the twentieth century has focused almost exclusively on finding the neuroanatomical (i.e., structural) and the neuro-physiological (i.e., activational) correlates of perceptual properties that are purely *natural* or *physical* in nature. The central and programmatic question has been as follows. Where in the brain, and by what processes, do we recognize such properties as color, shape, motion, sound, taste, aroma, temperature, texture, bodily damage, relative distance, and so on? The pursuit of such questions has led to real insights, and we have long been able to provide a map of the various areas in the brain that seem centrally involved in each of the functions mentioned.

The discovery technique is simple in concept. Insert a long, thin micro-electrode into any one of the cells in the cortical area in question (the brain has no pain sensors, so the experimental animal is utterly unaware of this telephone tap), and then see whether and how that cell responds when the animal is shown color or motion, or hears tones, or feels warmth and cold, and so on. In this fashion, a functional map is painstakingly produced. Figure 5.4 provides a quick look at the several primary and secondary sensory cortices and their positions within the rear half of a typical primate cerebral cortex.

But what about the front half of the cortex, the so-called frontal lobe? What is it for? The conventional but vague answer is, "to formulate potential motor behaviors for delivery to and execution by the motor cortex." Here we possess much less insight into the significance of these cortical structures and their neuronal activities. We cannot manipulate the input to those areas in any detail, as we can with the several sensory areas, because the input received by the premotor areas comes ultimately from all over the brain: areas that are already high up in the processing hierarchy, areas a long way from the sensory periphery where we can easily control what is and is not presented.

On the other hand, we can insert microelectrodes as before, but this time stimulate the target cell rather than record from it. In the motor cortex itself, this works beautifully. If we briefly stimulate the cells in certain areas, certain muscles in the body twitch, and there is a systematic correspondence between the areas of motor cortex and the muscles they control. In short, the motor strip itself constitutes a well-ordered map of the body's many muscles, much as the primary visual cortex is a map of the eye's retina. Stimulating single cells outside of and upstream from the motor areas, however, produces little or nothing in the way of behavioral response, presumably because the production of actual behavior requires smooth sequences of large activation vectors involving many thousands of cells at once. That kind of stimulation we still lack the technology to produce.

A conventional education in neuroscience thus leaves one wondering exactly how the entire spectrum of sensory inputs processed in the rear half of the brain finally gets transformed into some appropriate motor outputs formulated in the front half of the brain. This is indeed a genuine problem, and it is no wonder that researchers have found it so difficult. From the perspective we have gained from our study of artificial networks, we can see how complex the business of vector coding and vector transformation must be in something as large as the brain.

Plainly, sleuthing out the brain's complete sensorimotor strategy would be a daunting task even if the brain were an artificial network, one whose every synaptic connection strength were known and all of whose neuronal activation levels were open to continuous and simultaneous monitoring. But a living brain is not so accommodating. Its connection strengths are mostly inaccessible, and monitoring the activity of more than a few cells at a time is currently impossible.

This is one of the reasons why the recent artificial network models have made possible so much progress. We can learn things from the models that we might never have learned from the brain directly. And we can then return to the biological brain with some new and better-informed experimental questions to pose—questions concerning the empirical faithfulness of our network models, questions that we do have some hope of answering. Accordingly, the hidden transformations that produce behavior from perceptual input need not remain hidden after all.

If we aspire to track them down, however, we need to broaden our conception of the problem. In particular, we should be wary of the assumption that perception is first and foremost the perception of purely physical features in the world. And we should be wary of the correlative assumption that behavioral output is first and primarily the manipulation of physical objects.

We should be wary because we already know that humans and other social animals are keenly sensitive, perceptually, to *social* features of their surroundings. And because we already know that humans and social animals manipulate their *social* environment as well as their purely physical surroundings. And above all, because we already know that infants in most social species begin acquiring their *social* coordination at least as early as they begin learning sensorimotor coordination in its purely physical sense. Even infants can discriminate a smile from a scowl, a kind tone of voice from a hostile tone, a humorous exchange from a fractious one. And even an infant can successfully call for protection, induce feeding behavior, and invite affection and play.

I do not mean to suggest that social properties are anything more, ultimately, than just intricate aspects of the purely physical world. Nor do I wish to suggest that they have independent causal properties over and above what is captured by physics and chemistry. What I do wish to assert is that, in learning to represent the world, the brains of infant social creatures focus naturally and relentlessly on the social features of their local environment, often slighting physical features that will later seem unmissable. Human children, for example, typically do not acquire command of the basic color vocabulary until their third or fourth year of life, long after they have gained linguistic competence on matters such as anger, promises, friendship, ownership, and love. As a parent, I was quite surprised to discover this in my own children, and surprised again to learn that the pattern is quite general. But perhaps I should not have been. The social features listed are far more important to a young child's practical life than are the endlessly various colors.

The general lesson is plain. As social infants partition their activation spaces, the categories that form are just as often social categories as they are natural or physical categories. In apportioning neuronal resources for important cognitive tasks, the brain expends roughly as much of those resources on representing and controlling social reality as it does on representing and controlling physical reality.

Look once again, in the light of these remarks, at the brain in figure 5.3. Notice the unmapped frontal half and the large, unmapped areas of the rear half. Might some of these areas be principally involved in *social* perception and action? Might they be teeming with vast vectorial sequences representing *social* realities of one sort or other? Indeed, once the question is raised, why stop with these areas? Might the so-called primary sensory cortical areas—for touch, vision, and hearing especially—be as much in the business of grasping and processing social facts as they are in the business of grasping and processing purely physical facts? These two functions are certainly not mutually exclusive.

I think the answer is almost certainly yes to all of these questions. We lack intricate brain maps for social features comparable to existing brain maps for physical features, not because they are nonexistent, I suggest, but rather because we have not looked for them with a determination comparable to the physical case.

Moral Perception and Moral Understanding

Although there is no room to detail the case here, an examination of how neural networks sustain scientific understanding reveals that the role of learned *prototypes* and their continual redeployment in new domains of phenomena is central to the scientific process (Kuhn 1962; Churchland 1989, 1995). Specific *rules* or "laws of nature" play an undeniably important but nonetheless secondary role, mostly in the social business of communicating or teaching scientific skills. One's scientific understanding is lodged primarily in one's acquired hierarchy of structural and dynamical prototypes, not primarily in a set of linguistic formulas.

In a parallel fashion, neural network research has revealed how our knowledge of a language may be embodied in a hierarchy of *prototypes* for verbal sequences that admit of varied instances and indefinitely many combinations, rather than in a set of specific rules-to-be-followed (Elman 1992). Of course, we can and do state grammatical rules, but a child's grammatical competence in no way depends on ever hearing them uttered or being able to state them. It may be that the main function of such rules resides in the social business of describing and refining our linguistic skills. One's grammatical capacity, at its original core, may consist of something other than a list of internalized rules-to-be-followed.

With these two points in mind, let us finally turn to the celebrated matter of our *moral* capacity. Let us address our ability to recognize cruelty and kindness, avarice and generosity, treachery and honor, mendacity and honesty, the Cowardly Way Out and the Right Thing to Do. Here, once again, the intellectual tradition of Western moral philosophy is focused on *rules*, on specific laws or principles. These are supposed to

govern one's behavior, to the extent that one's behavior is moral at all. ,
And the discussion has always centered on which rules are the truly valid,
correct, or binding rules.

I have no wish whatever to minimize the importance of that ongoing
moral conversation. It is an essential part of humanity's collective cognitive adventure, and I would be honored to make even the most modest of
contributions to it. Nevertheless, it may be that a normal human's capacity ,
for moral perception, cognition, deliberation, and action has rather less to
do with rules, whether internal or external, than is commonly supposed. ,

What is the alternative to a rule-based account of our moral capacity?
The alternative is a hierarchy of learned prototypes, for both moral perception and moral behavior, prototypes embodied in the well-tuned configu-
ration of a neural network's synaptic weights. We may here find a more
fruitful path to understanding the nature of *moral learning, moral insight,
moral disagreements, moral failings, moral pathologies,* and *moral growth* at the
level of entire societies. Let us explore this alternative, just to see how
some familiar territory looks from a new and different hilltop.

One of the lessons of neural network research is that one's capacity
for recognizing and discriminating perceptual properties usually outstrips
one's ability to articulate or express the basis of such discriminations in
words. Tastes and colors are the leading examples, but the point quickly
shows itself to have a much broader application. Faces, too, are something
we can discriminate, recognize, and remember to a degree that exceeds
any verbal articulation we could possibly provide. The facial expression of
emotions is evidently a third example. The recognition of sounds is a
fourth. In fact, the cognitive priority of the preverbal over the verbal
shows itself, upon examination, to be a feature of almost all of our cognitive categories.

This supraverbal grasp of the world's many dimensions of variation is
perhaps the main point of having concepts: it allows us to deal appropriately with the always novel but never *entirely* novel situations flowing
endlessly toward us from an open-ended future. That same flexible readiness characterizes our social and moral concepts no less than our physical
concepts. And our moral concepts show the same penetration and supraverbal sophistication shown by nonmoral concepts. One's ability to
recognize instances of cruelty, patience, meanness, and courage, for instance, far outstrips one's capacity for verbal definition of those notions.
One's diffuse expectations of their likely consequences similarly exceed
any verbal formulas that one could offer or construct, and those expectations are much the more penetrating because of it. All told, moral cognition would seem to display the same profile or signature that in other
domains indicates the activity of a well-tuned neural network underlying
the whole process.

Figure 5.5
The old woman/young woman, a classic case of a visually ambiguous figure. The old woman is looking left and slightly toward us with her chin buried in her ruff. The young woman is looking to the left but away from us; her nose is barely visible, but her left ear, jawline, and choker necklace are directly before us.

If this is so, then moral perception will be subject to same ambiguities that characterize perception generally. Moral perception will be subject to the same modulation, shaping, and occasional "prejudice" that recurrent pathways make possible. By the same token, moral perception will occasionally be capable of the same cognitive "reversals" that we see in such examples as the old/young woman in figure 5.5. Pursuing the parallel further, it should also display cases where one's first moral reaction to a novel social situation is simply moral confusion, but where a little background knowledge or collateral information suddenly resolves that confusion into an example of something familiar, into an unexpected instance of some familiar moral prototype.

On these same assumptions, moral learning will be a matter of slowly generating a hierarchy of moral prototypes, presumably from a substantial number of relevant *examples* of the moral kinds at issue. Hence the relevance of stories and fables, and, above all, the ongoing relevance of the parental example of interpersonal behavior, and parental commentary on and consistent guidance of childhood behavior. No child can learn the route to love and laughter entirely unaided, and no child will escape the pitfalls of selfishness and chronic conflict without an environment filled with examples to the contrary.

People with moral perception will be people who have learned those lessons well. People with reliable moral perception will be those who can protect their moral perception from the predations of self-deception and the corruptions of self-service. And, let us add, from the predations of group-think and the corruptions of fanaticism, which involves a rapacious disrespect for the moral cognition of others.

People with unusually penetrating moral insight will be those who can see a problematic moral situation in more than one way, and who can evaluate the relative accuracy and relevance of those competing interpretations. Such people will be those with unusual moral *imagination*, and a critical capacity to match. The former virtue will require a rich library of moral prototypes from which to draw, and special skills in the recurrent manipulation of one's moral perception. The latter virtue will require a keen eye for local divergences from any presumptive prototype and a willingness to take them seriously as grounds for finding some alternative understanding. Such people will by definition be rare, although all of us have some moral imagination, and all of us some capacity for criticism.

Accordingly, moral disagreements will be less a matter of interpersonal conflict over what "moral rules" to follow and more a matter of interpersonal divergence as to what moral prototype best characterizes the situation at issue. It will be more a matter, that is, of divergences over what kind of case we are confronting in the first place. Moral argument and moral persuasion, on this view, will most typically be a matter of trying to make salient this, that, or the other feature of the problematic situation, in hopes of winning one's opponent's assent to the local appropriateness of one general moral prototype over another. A nonmoral parallel of this phenomenon can again be found in the old/young woman example of figure 5.5. If that figure were a photograph, say, and if there were some issue as to what it was really a picture of, I think we would agree that the young-woman interpretation is by far the more realistic of the two. The old-woman interpretation, by comparison, asks us to believe in the reality of a hyperbolic cartoon.

A genuinely moral example of this point about the nature of moral disagreement can be found in the current issue over a woman's right to abort a first-trimester pregnancy without legal impediment. One side of the debate considers the status of the early fetus and invokes the moral prototype of a Person, albeit a very tiny and incomplete person, a person who is defenseless for that very reason. The other side of the debate addresses the same situation and invokes the prototype of a tiny and possibly unwelcome Growth, as yet no more a person than is a cyst or a cluster of one's own skin cells. The first prototype bids us bring to bear all the presumptive rights of protection due any person, especially one that is young and defenseless. The second prototype bids us leave the woman to deal with the tiny growth as she sees fit, depending on the value it may or may not currently have for her, relative to her own long-term plans as an independently rightful human. Moral argument, in this case as elsewhere, typically consists in urging the accuracy or the poverty of the prototypes at issue as portrayals of the situation at hand.

I cite this example not to enter into this debate, nor to presume on the patience of either party. I cite it to illustrate a point about the nature of moral disagreements and the nature of moral arguments. The point is that real disagreements need not be and seldom are about what explicit moral rules are true or false. The adversaries in this case might even agree on the obvious principles lurking in the area, such as, "It is prima facie wrong to kill any person." The disagreement here lies at a level deeper than that glib anodyne. It lies in a disagreement about the boundaries of the category "person" and hence about whether the explicit principle even applies to the case at hand. It lies in a divergence in the way people perceive or interpret the social world they encounter and in their inevitably divergent behavioral responses to that world.

Whatever the eventual resolution of this divergence of moral cognition, it is antecedently plain that both parties to this debate are driven by some or other application of a moral prototype. But not all conflicts are thus morally grounded. Interpersonal conflicts are regularly no more principled than that between a jackal and a buzzard quarreling over a steaming carcass, or a pair of two-year-old human children screaming in frustration at a tug-of-war over the same toy. This returns us, naturally enough, to the matter of moral development in children and to the matter of the occasional failures of such development. How do such failures look, on the trained-network model here being explored?

Some of them recall a view from antiquity. Plato was inclined to argue, at least in his occasional voice as Socrates, that no man ever knowingly does wrong. For if he recognizes the action as being genuinely wrong—rather than just "thought to be wrong by others"—what motive could he possibly have to perform it? Generations of students have rejected Plato's suggestion, and rightly so. But Plato's point, however overstated, remains an instructive one: an enormous portion of human moral misbehavior is due primarily to cognitive failures of one kind or another.

Such failures are inevitable. We have neither infinite intelligence nor total information. No one is perfect. But some people, as we know, are notably less perfect than the norm, and their failures are systematic. In fact, some people are rightly judged to be chronic troublemakers, terminal narcissists, thoughtless blockheads, and treacherous snakes, not to mention bullies and sadists. Whence stem these sorry failings?

From many sources, no doubt. But we may note right at the beginning that a simple failure to develop the normal range of moral perception and social skills will account for a great deal here. Consider the child who, for whatever reasons, learns only very slowly to distinguish the minute-by-minute flux of rights, expectations, entitlements, and duties as they are created and canceled in the course of an afternoon at the day-care center, an outing with siblings, or a playground game of hide and seek. Such a child is doomed to chronic conflict with other children—doomed to cause them disappointment, frustration, and eventually anger, all of it directed at him.

Moreover, he has all of it coming, despite the fact that a flinty-eyed determination to "flout the rules" is *not* what lies behind his unacceptable behavior. The child is a moral cretin because he has not acquired the skills already flourishing in the others. He is missing skills of recognition to begin with, and also the skills of matching his behavior to the moral circumstance at hand, even when it is dimly recognized. The child steps out of turn, seizes disallowed advantages, reacts badly to constraints binding on everyone, denies earned approval to others, and is blind to opportunities for profitable cooperation. His failure to develop and deploy a roughly normal hierarchy of social and moral prototypes may seem tragic, and it is. But one's sympathies must lie with the other children when, after due patience runs out, they drive the miscreant howling from the playground.

What holds for a playground community holds for adult communities as well. We all know adult humans whose behavior recalls to some degree the bleak portrait just outlined. They are, to put the point gently, unskilled in social practices. Moreover, all of them pay a stiff and continuing price for their failure. Overt retribution aside, they miss out on the profound and ever-compounding advantages that successful socialization brings, specifically, the intricate practical, cognitive, and emotional commerce that lifts everyone in its embrace.

The Basis of Moral Character

This quick portrait of the moral miscreant invites a correspondingly altered portrait of the morally successful person. The common picture of the Moral Man as one who has acquiesced in a set of explicit rules imposed from the outside—from God, perhaps, or from society—is dubious in the extreme. A relentless commitment to a handful of explicit rules does not make one a morally successful or a morally insightful person. That is the path of the Bible Thumper and the Waver of Mao's *Little Red Book*. The price of virtue is a good deal higher than that, and the path thereto a good deal longer. It is much more accurate to see the moral person as one who has acquired a complex set of subtle and enviable skills: perceptual, cognitive, and behavioral.

This was, of course, the view of Aristotle, to recall another name from antiquity. Moral virtue, as he saw it, was something acquired and refined over a lifetime of social experience, not something swallowed whole from an outside authority. It was a matter of developing a set of largely inarticulable skills, a matter of *practical* wisdom. Aristotle's perspective and the neural network perspective here converge.

To see this more clearly, focus now on the single individual, one who grows up among creatures with a more or less common human nature, in an environment of ongoing social practices and presumptive moral wisdom already in place. The child's initiation into that smooth collective practice takes time—time to learn how to recognize a large variety of prototypical social situations, time to learn how to deal with those situations, time to learn how to balance or arbitrate conflicting perceptions and conflicting demands, and time to learn the sorts of patience and self-control that characterize mature skills in any domain of activity. After all, there is nothing essentially moral about learning to defer immediate gratification in favor of later or more diffuse rewards.

So far as the child's brain is concerned, such learning, such neural representation, and such deployment of those prototypical resources are all indistinguishable from their counterparts in the acquisition of skills generally. There are real successes, real failures, real confusions, and real rewards in the long-term quality of life that one's moral skills produce. As in the case of internalizing mankind's scientific knowledge, a person who internalizes mankind's moral knowledge is a more powerful, effective, and resourceful creature because of it. To draw the parallels here drawn is to emphasize the practical or pragmatic nature of both scientific and broadly normative knowledge. It is to emphasize the fact that both embody different forms of know-how: how to navigate the natural world in the former case and how to navigate the social world in the latter.

This portrait of the moral person as one who has acquired a certain family of perceptual and behavioral *skills* contrasts sharply with the more traditional accounts that picture the moral person as one who has agreed to follow a certain set of *rules* (for example, "Always keep your promises") or alternatively, as one who has a certain set of overriding *desires* (to maximize the general happiness). Both of these more traditional accounts are badly out of focus.

For one thing, it is just not possible to capture, in a set of explicit imperative sentences or rules, more than a small part of the practical wisdom possessed by a mature moral individual. It is no more possible here than in the case of any other form of expertise—scientific, athletic, technological, artistic, or political. The sheer amount of information stored in a well-trained network the size of a human brain, and the massively distributed and exquisitely context-sensitive ways in which it is stored therein, preclude its complete expression in a handful of sentences, or even a large bookful. Statable rules are not the *basis* of one's moral character.

They are merely its pale and partial reflection at the comparatively impotent level of language.

If rules do not do it, neither are suitable desires the true basis of anyone's moral character. Certainly they are not sufficient. A person might have an all-consuming desire to maximize human happiness. But if that person has no comprehension of what sorts of things genuinely serve lasting human happiness; no capacity for recognizing other people's emotions, aspirations, and current purposes; no ability to engage in smoothly cooperative undertakings; no skills whatever at pursuing that all-consuming desire; then that person is not a moral saint. He is a pathetic fool, a hopeless busybody, a loose cannon, and a serious menace to his society.

Neither are canonical desires obviously necessary. A man may have, as his most basic and overriding desire in life, the desire to see his own children mature and prosper. To him, let us suppose, everything else is distantly secondary. And yet such a person may still be numbered among the most consummately moral people of his community, so long as he pursues his personal goal, as others may pursue theirs, in a fashion that is scrupulously fair to the aspirations of others and ever protective of the practices that serve everyone's aspirations indifferently.

Attempting to portray either accepted rules or canonical desires as the basis of moral character has the further disadvantage of inviting the skeptic's hostile question: "Why should I follow those rules?" in the first case, and "What if I don't have those desires?" in the second. If, however, we reconceive strong moral character as the possession of a broad family of perceptual, cognitive, and behavioral skills in the social domain, then the skeptic's question must become, "Why should I acquire those skills?" To which the honest answer is, "Because they are easily the most important skills you will ever learn."

This novel perspective on the nature of human cognition, both scientific and moral, comes to us from two disciplines—cognitive neuroscience and connectionist artificial intelligence—that had no prior interest in or connection with either the philosophy of science or moral theory. And yet the impact on both these philosophical disciplines is destined to be revolutionary. Not because an understanding of neural networks will obviate the task of scientists or of moral-political philosophers. Not for a second. Substantive science and substantive ethics will still have to be done, by scientists and by moralists and mostly in the empirical trenches. Rather, what will change is our conception of the nature of scientific and moral knowledge, as it lives and breathes within the brains of real creatures. Thus, the impact on *meta*ethics is modestly obvious already. No doubt a revolution in moral psychology will eventually have some impact on substantive ethics as well —on matters of moral training, moral pathology, and moral correction, for example. But that is for moral philosophers to work through, not cognitive theorists. The message of this essay is that an ongoing conversation between these two communities has now become essential.

Acknowledgment

This essay is excerpted from chapters 6 and 10 of P. M. Churchland, *The Engine of Reason, the Seat of the Soul: A Philosophical Journey into the Brain*, Cambridge, Mass., 1995. I thank The MIT Press for permission to print some of that material here.

References

Churchland, P. M. (1989). *A Neurocomputational Perspective: The Nature of Mind and the Structure of Science*. Cambridge, Mass.: Bradford Books/MIT Press.

Churchland, P. M. 1995. *The Engine of Reason, the Seat of the Soul: A Philosophical Journey into the Brain*. Cambridge, Mass.: Bradford Books/MIT Press.

Elman, J. L. 1992. "Grammatical Structure and Distributed Representations." In S. Davis, ed., *Connectionism: Theory and Practice*. Oxford: Oxford University Press.

Kuhn, T. S. 1962. *The Structure of Scientific Revolutions*. Chicago: University of Chicago Press.

Chapter 6
Connectionism, Moral Cognition, and Collaborative Problem Solving
Andy Clark

How should linguistically formulated moral principles figure in an account of our moral understanding and practice? Do such principles lie at the very heart of moral reason (Kohlberg 1981)? Or do they constitute only a shallow, distortive gloss on a richer prototype-based moral understanding (Churchland 1989; Dreyfus and Dreyfus 1990)? The latter view has recently gained currency as part of a wider reassessment of the proper cognitive scientific image of human cognition—a reassessment rooted in the successes of a class of computational approaches known as connectionist, parallel distributed processing, or neural network models (McClelland, Rumelhart, and the PDP Research Group 1986). In this chapter I will argue that such approaches call not for the marginalization of the role of linguistically formulated moral rules and principles but for a thorough reconception of that role. This reconception reveals such summary maxims as the guides and signposts that enable collaborative moral exploration rather than as failed attempts to capture the rich structure of our individual moral knowledge. The force of this reconception eludes us, however, if we cast public linguistic exchange as primarily a tool for manipulating the moral understanding of other agents (Churchland 1989). Instead, we must focus on the role of such exchanges in attempts to engage in genuinely collaborative moral problem solving. A satisfying connectionist model of moral cognition will need to address the additional inner mechanisms by which such collaborative activity becomes possible and to recognize the ways in which such activity transforms the space of our moral possibilities.

Connectionism: From Rules to Prototypes

There is a conception of moral reason that informed many classical philosophical treatments but that has recently been called into question. At the heart of this conception lies a vision of informed moral choice as involving the isolation and application of an appropriate law or rule. Thus, to give a simplistic example, we might, on encountering some complex social situation, see it as falling under a rule prohibiting lying and act accordingly. Of course, the body of rules we are supposed (on this conception) to have internalized need not be so simple. Such a moral code could, as Ruth Barcan Marcus points out, be highly elaborate.[1] Principles could be hedged by exception clauses or rank ordered to help deal with potential conflicts. The logic of the moral code could be a fuzzy logic, or a nonmonotonic one, or something else. But however elaborate, the basic vision remains the same. It is a vision in which moral judgments are taken to involve "the judging of particular cases as falling under a particular moral concept, and thereby being governed by a specific moral rule" (Johnson 1993, 207). Johnson calls such a doctrine "moral law folk theory" (4–6). This doctrine, he claims, permeates our cultural heritage and hence underpins both lay and philosophical conceptions of the moral life. Yet it is a doctrine that, he argues, is radically mistaken, and indeed morally incorrect. It is, Johnson claims, "morally irresponsible to think and act as though we possess a universal, disembodied reason that generates absolute rules, decision-making procedures, and universal or categorical laws" (5).

Moral law folk theory is false, Johnson suggests, because it "presupposes a false account of the nature of human concepts and reason." This false account depicts concepts (including those that figure in putative moral rules and universals) as possessing classical structure. A concept possesses classical structure if it can be unpacked so as to reveal a set of necessary and sufficient conditions for its application. These necessary and sufficient conditions would effectively define the concept. To grasp the concept, on such a model, is to know that definition and to rely on it when called upon to deploy the concept. The trouble, as is now well known (Smith and Medin 1981; Rosch 1973) is that most (perhaps all) human concepts do not possess classical structure. Thus, whereas the classical model predicts that instances should fall squarely within or outside the scope of a given concept (according to whether the necessary and sufficient conditions are or are not met), robust experimental results reveal strong so-called typicality effects. Instances are classified as more or less falling under a concept or category according, it seems, to the perceived distance of the instance from prototypical cases. Thus, a dog is considered a better example of a pet than a tortoise, and a robin a better example of a bird than a pigeon. Such findings sit uneasily beside the classical image.[2] It seems we do not simply test for the presence or absence of a neat list of defining features and judge the concept applicable or inapplicable accordingly. In place of definitions and application rules invoking them, we face a vision of cognition organized around prototypical instances. To bring this vision into focus, we need to say a little more about the very idea of a prototype.

"Prototype" is sometimes used to mean merely a stereotypic example of the membership of some category. Thus understood, the stereotypic pet might be a dog, or the stereotypic crime a robbery. The recent popularity of prototype-invoking accounts in psychology and artificial intelligence, however, depends on a related but importantly different conception. Here, the notion of prototype is not the notion of a real, concrete exemplar. Rather, it is the notion of the statistical central tendency of a body of concrete exemplars. Such central tendency is calculated by treating each concrete example as a set of co-occuring features and generating (as the prototype) a kind of artificial exemplar that combines the statistically most salient features. Thus, the prototypical pet may include both dog and cat features, and the prototypical crime may include both harm to the person and loss of property. Concrete exemplars and rich worldly experience are still crucial, but they act as sources of data from which these artificial prototypes are constructed. Novel cases are then judged to fall under a concept (such as "pet" or "crime") according to the distance separating the set of features they exhibit from the prototypical feature complex—hence, the typicality effects mentioned earlier. (For a full review, see Clark 1993.)

Such a vision of prototype-based reason fits very satisfyingly with a particular model of information storage in the brain. This is the model of state-space representation, which draws on both neuroscientific conjecture and recent work with computational simulations of a broadly connectionist type (P. S. Churchland and T. J. Sejnowski 1992; P. M. Churchland 1989; McClelland, Rumelhart, and the PDP Research Group 1986, vols. 1, 2; Clark 1989, 1993). The main philosophical proponent of the state-space conception is undoubtedly Paul Churchland, who has also urged its impor- tance for conceptions of moral reason (P. M. Churchland 1989, chap. 14). The flavor of the state-space approach is best conveyed by tracing out a simple example.

Consider the brain's representation of color. This representation (the example comes from P. M. Churchland 1989, 104) is fruitfully conceived as involving a three-dimensional (3D) state space (Land's color cube; see Land 1977) in which the dimensions (axes) reflect (1) long-wave reflectance, (2) medium-wave reflectance, and (3) short-wave reflectance. Each such dimension, Churchland conjectures, may correspond to the ac- tivity of downstream neural groups tuned to the activity of three different kinds of retinal cone. Within such a 3D space, white and black occupy diametrically opposed locations, while red and orange are quite close together. Our judgments concerning the perceived similarity-difference relations between colors may thus be explained as reflecting distance in this color-state space. The space thus exhibits what has been termed (Clark 1989, 1993) an inbuilt semantic metric.

Connectionist networks constitute one way of both implementing and acquiring representational spaces of this sort. Such networks consist of a complex of units (simple processing elements) and connections. The connections may be positive or negative valued (excitatory or inhibitory). The features of a stimulus are coded as numerical values (corresponding to the intensity of the feature or the degree to which it is present) across a designated group of units. These values are then differentially propagated through the network courtesy of the positive or negative connection weights. A good assignment of weights is one that ensures that activity in designated output units corresponds to some target input-output function.[3] Several layers of units may intervene between input and output. The activity of the units in each such layer will generally correspond to some kind of recoding of the input data that simplifies further processing. It is often fruitful to take each unit of such an intervening ("hidden") layer as one dimension of an acquired representational state space and to investigate the way the system responds to specific inputs by creating patterns of activity that define locations in this space (hidden unit activation space).

The great achievement of connectionism is to have discovered a set of learning rules that enables such systems to find their own assignments of weights to connections. The operation of such learning rules (which I shall not attempt to describe here, but see Clark 1989 and Churchland 1989 for accessible treatments) results in the construction of high-dimensional state spaces with associated semantic metrics. Four important features of this constructive process are:

1. It is exemplar-driven.
2. It is not bound by the similarity metric on the input vector.
3. It yields prototype-style representations.
4. It treats inference and reasoning as vectorial transformations across state spaces.

The learning process is exemplar driven in that the tuning of the weights is achieved by exposure to concrete cases. Thus, a network whose target is to transform written text into coding for phonemes (hence speech) does not have its weights changed by exposure to rules of text-phoneme conversion. Instead, it must be exposed to textual inputs, allowed to output an attempted coding for phonemes, and then amend its weights according to the difference between the target output and its actual performance.

The weight assignments that a net thus acquires can exploit hidden layers so as to dilate and compress the input space. Thus, two exemplars whose coding at the input layer is very similar may be pulled apart by the weights leading up the hidden layer. This is useful if, for example, two visual input descriptions are quite similar yet require very different responses. Thus, two situations that are visually very similar (such as a person giving money to a beggar versus the same person giving money to a mugger holding a knife) may require very different responses. In such cases, the net can learn to use the hidden layer to recode the inputs in a new way such that the pattern of hidden unit activation is radically dissimilar in the cases just described. Correlatively, it may learn to code superficially dissimilar cases (such as giving money to a beggar and posting a check to a charity) with very similar hidden unit patterns. The state space defined by the hidden units might thus come to reflect a moral metric, whereas the input space depicted a visual one.

Within such state spaces, the mode of representation will come to exhibit features of prototype-style encodings. The reason is that features common to several training examples will figure in more episodes of weight adjustment than the less common features. As a result, the system will become especially adept at encoding and responding to such features. In addition, the learning regime will ensure that features that commonly occur together in the exemplars become strongly mutually associated. The upshot is that the system extracts the so-called central tendency of the body of exemplars, that is, a complex of common, co-occurring features. Moreover, multiple such complexes can be extracted and stored by a single net. McClelland and Rumelhart (1976) describe a net that (1) learns to recognize individual dogs by associating visual information with names, (2) extracts the central tendency of the body of dog exemplars, and hence exhibits knowledge of a prototypical set of dog features, and (3) can perform this trick for several different categories, simultaneously encoding knowledge about dogs, cats, and bagels in a single network. These various prototypes (of dog, cat, and bagel) are each coded for by distinct patterns of unit activation and hence determine different locations in a general state space whose dimensionality corresponds to the number of processing units. Individual dogs are coded by points relatively close to the dog prototype. Dogs and cats share more features with each other than do either with bagels; hence, the dog and cat prototypes lie relatively close together and at some fair remove from the bagel prototype. The system can use its knowledge of prototypical feature complexes to behave sensibly given novel inputs. To the extent that some new input exhibits several familiar features (for example, a half-dog/half-cat) the system will assign it to an appropriate location (in this case, midway between the dog and cat prototypes) and hence yield suitable outputs.

Reasoning and inference can now be reconstructed as processes of pattern completion and pattern extension. A network exposed to an input depicting the visual features of a red-spotted face may learn to activate a pattern of hidden units corresponding to a diagnosis of measles and a prescription of penicillin. The vector-to-vector transformation involved is of a piece with that by which we perform simple acts of recognition and categorization such as naming a familiar dog. On the face of it, it is a million miles away from the intellectualist artificial intelligence model, which would have us consult a body of rules and principles and issue a medical judgment accordingly. (For a variety of similar examples and claims, see P. M. Churchland 1989, chap. 10.)

With this rough understanding of vector transformation models in place, we can now begin to address the issues concerning moral knowledge and reason. The primary lessons of the new approach, when applied to the moral domain, look to be twofold. First, the successful acquisition of moral knowledge may be heavily dependent on exposure not to abstractly formulated rules and principles but to concrete examples of moral judgment and behavior. (Literature, by depicting complex moral situations, may be seen as another kind of concrete case—virtual moral reality, if you will.) Second, our individual moral knowledge and reasoning may not be fully reconstructible in the linguistic space afforded by public language moral dialogue and discussion. This will be the case if, as seems likely, the internal representational space (or spaces) involved has even a fairly modest number of dimensions. Our sense of smell, as P. M. Churchland (1989, 106) notes, looks to involve at least a 6D space. If each dimension can take just ten different values, a space of 10^6 distinct locations is immediately available. Dog olfactory space, Churchland calculates, is of the order of 30^7 (22 billion) possible locations (compare to a world population of just 3.5 billion). State-space encoding thus allows even limited internal resources of units and weights to support representational spaces of great magnitude. Given the size of the brain's resources, the expressive capacity of biologically realistic inner systems looks unimaginably huge. The attempt to condense the moral expertise encoded by such a system into a set of rules and principles that can be economically expressed by a few sentences of public language may thus be wildly optimistic, akin to trying to reduce a dog's olfactory skills to a small body of prose.

These two implications (concerning the role of exemplars and the resistance of moral knowledge to summary linguistic expression) are remarked on by several recent writers.[4] Goldman (1993) notes the central role of exemplars, Churchland (1989) stresses in addition the general resistance of high-dimensional state-space encoded knowledge to linguistic expression, and Johnson (1993) describes how prototype-style encodings take us "beyond rules." The rule-based moral vision, according to this emerging consensus, is a doomed attempt to reconstruct the high-dimensional space of moral reason in a fundamentally low-dimensional medium. Such a diagnosis casts valuable light on questions concerning the rationality of moral thought. A well-tuned network, in command of state spaces of great complexity, may issue judgments that are by no means irrational but yet resist quasi-deductive linguistic reconstruction as the conclusion of some moral argument that takes summary expressions of moral rules and principles as its premises. Such a vision is by no means new. Nagel comments that "the fact that one cannot say why a certain decision is the correct one ... does not mean that the claim to correctness is meaningless.... What makes this possible is *judgement* [which can] in many cases be relied on to take up the slack that remains beyond the limits of explicit rational argument" (Nagel 1987, 180), or again, "We know what is right in a particular case by what we may call an immediate judgement, or an intuitive subsumption ... moral judgements are not discursive" (Bradley 1876). As Bradley points out, phrases like "judgment" and "intuitive subsumption" are "perhaps not very luminous" (65). The value of the cognitive scientific excursion into state-space representation is just to cast a little light. It helps make concrete sense of a form of rational moral choice that nonetheless outruns what Nagel called "explicit rational argument."

The realization that individual moral know-how may resist expression in the form of any set of summary moral rules and principles is important. But it has mistakenly (or so I shall argue) led some writers to marginalize the role of such summary linguistic expressions in our moral life. It is this correlative marginalization that I now set out to resist. Such marginalization, I shall suggest, is the result of a common error: the error of seeing talk in general (and moral talk in particular) as primarily the attempt to reflect the fine-grained contents of individual thought. Such a view seems implicit in, for example, Paul Churchland's general skepticism concerning linguistic renditions. Such skepticism is evidenced in passages such as the following: "Any declarative sentence to which a speaker would give confident assent is merely a one-dimensional projection—through the compound lens of Wernicke's and Broca's areas onto the idiosyncratic surface of the speaker's language—a one-dimensional projection of a [high] dimensional solid that is an element in his true kinematical state" (P. M. Churchland 1989, 18). The high-dimensional solid is, of course, the internalized prototype-style know-how contained in a trained-up neural network. It is this know-how whose linguistic echo is but the flickering shadow on the wall of Plato's cave (Churchland 1989, 18).

More radically still, Dreyfus and Dreyfus (1990) go so far as to demote such low-dimensional linguaform projections to the status of mere tools for the novice—ladders to be kicked away by the true moral expert. Once again, the radical claim has some plausible roots in the observation that (as far as we can tell) truly expert ability (at chess, car driving, philosophy, moral reasoning) is not subserved by a set of compact rules or principles encoded quasi-linguistically by the brain. Instead, it is subserved by the operation of a fast, unreflective, connectionist-style resource or resources whose operation yields "everyday, intuitive ethical expertise" (246).

According to Dreyfus and Dreyfus, the expert, under normal conditions, "does not *deliberate*. She does not reason. She does not even act deliberately. She simply spontaneously does what has normally worked and, naturally, it normally works" (243).

This kind of fluid expertise comes, if it comes at all, only at the end of an extended learning history whose early stages are indeed marked by episodes of linguistic instruction. Dreyfus and Dreyfus in fact distinguish four stages that they claim precede fluid expertise: novice, advanced beginner, competence, and proficiency. Linguistic instruction figures prominently (unsurprisingly) in the initial novice stage, while linguistic reflection figures to a degree in all the other nonexpert stages: "The instruction process begins with the instructor decomposing the task environment into context-free features which the beginner can recognize without benefit of experience" (240). These context-free features are then used as components of rough-and-ready rules. Thus the would-be chess player is taught the numerical values of pieces (in context-free terms) and told to exchange pieces whenever a profit would accrue. Similarly, the would-be moral agent is told that to say intentionally what is false is to lie and that lying is in general to be avoided. (See Flanagan 1991 for a richer and more realistic treatment of this example.)

Dreyfus and Dreyfus are surely right to stress the role of language in novice learning. But they go on, wrongly I believe, to marginalize the role of language in truly expert behavior. They write that "principles and theories serve only for early stages of learning," and as a result, "the skill development model we are proposing ... demotes rational, post-conventional moral activity to the status of a regression to a pre-expert stage of moral development" (252–256).[5] We are thus invited to treat linguistic justification and linguistically couched reflection as mere beginners' tools—rough instruments not to be found in the tool kit of the true moral expert. I do not think this is the case. Rather, linguistic reflection and exchange enables a tuning and orchestration of moral response that is vital to moral expertise. What is needed is not a rejection of the role of summary linguistic expression and linguaform exchange in advanced moral cognition. Rather, we must reconceive that role. Such a reconception will occupy us for the remainder of this treatment.

Language as a Manipulative Tool

As a first move toward such a reconception, consider a question recently raised by P. M. Churchland. How, on a connectionist/prototype-based view does moral knowledge, once achieved, get modified and altered? Churchland bids us distinguish between the kinds of slow, experientially driven adaptive change and learning that configure the weights of individual networks over time and the kinds of fast flash-of-insight style "learning" that seems to occur when we suddenly see that a question with which we have been wrestling is easily solved once we reconceive the domain in the light of some new idea. The question Churchland raises is: How (if at all) is slow, connectionist-style learning to cope with fast flash-of-insight style conceptual change? The answer he develops is that it is able to do so because of the operation of so-called context fixers—additional inputs that are given alongside the regular input and that may cause an input that (alone) could not activate an existing prototype to in fact do so. Churchland terms such a process "conceptual redeployment" because it often leads to the reinvocation of prototypes developed in one domain in another, superficially very different, one.

Imagine someone trying to solve a problem. To solve it, if the approach outlined in the previous section is correct, is to activate an appropriate explanatory prototype. Sometimes, however, our attempts to access a satisfying (explanatory) prototype fail. One diagnosis may be that we do not command any appropriate prototype, in which case there is no alternative to slow, experience-based learning. But an alternative possibility is that we do command just such a prototype but have so far not called it up. This is where a good piece of context fixing can help. The idea is that a bare input that previously led to the activation of no fully explanatory prototype may suddenly, in the context of additional information, give rise to the activation of a developed and satisfying prototype by being led to exploit resources originally developed for a different purpose. Huygens, we are told, commanded a powerful wave prototype developed for water and sound media. Once he was led (by luck, scholarship, or something else) to combine questions regarding optics with context-fixing inputs concerning light, the optical questions were able to activate the rich and explanatory wave prototypes originally devised for the water domain. The conceptual revolution thus achieved did not involve slow, weight-adjustment-style learning, but rather consisted in "the unusual deployment of old resources" (23). The context-fixing information thus biases the treatment of an input vector in ways that can radically alter the prototype-invoking response of on-board, trained-up networks.

Now this, as Churchland notes, invites a certain perspective on linguaform debate, for linguistic exchanges can be seen as a means of providing fast, highly focused, context-fixing information. Such information may, as we have seen, induce others to activate prototypes they already command in situations in which those very prototypes would otherwise remain dormant. According to this view, moral debate does not work by attempting to trace out nomological-deductive arguments predicated on neat linguaform axioms. But summary moral rules and linguistic exchanges may nonetheless serve as context-fixing descriptions that prompt others to activate certain stored prototypes in preference to others (see, e.g., comments in Churchland 1989, 300). Applying our story to an example from Johnson (1993), a moral debate may consist in the exchange of context fixers, some of which push us toward activation of an "invasion of privacy" prototype while others prompt us to conceptualize the very same situation in terms of a "prevention of espionage" prototype.

Note that according to such a vision the linguaform expressions do not aim to embody the reasoning that underlies individual moral judgment. Instead, they figure in exchanges whose goal is simply to prompt another's rich prototype-based knowledge to settle on one existing prototype rather than another. Thus, talk of "unborn children" may bias prototype-activation one way, while talk of "unwanted pregnancy" may bias it another. Moral rules and principles, on this account, are nothing more than one possible kind of context-fixing input among many. Others could include well-chosen images or non-rule-invoking discourse. Thus understood, language simply provides one fast and flexible means of manipulating activity within already developed prototype spaces. It is a simple matter, however, to extend this treatment to encompass a special role for summary principles, etc. in individual moral reflection. To see how, consider a nonmoral example case.

Kirsh and Maglio (1992, 1994) have investigated the roles of reaction and reflection in expert performance of the computer game Tetris in which the player attempts to accumulate a high score by the compact placement of geometric objects (Tetrazoids, or just Zoids) that fall down from the top of the screen. As a Zoid descends, the player can manipulate its fall by rotating it, moving it to the right or left, or instantly relocating it at the resting point of its current trajectory. When a Zoid comes to rest, a new one appears at the top of the screen. The speed of fall increases with score, and (the saving grace) a full row (one in which each screen location is filled by a Zoid) disappears entirely. When the player falls behind in Zoid placement and the screen fills up so that new Zoids cannot enter it, the game ends. Advanced play thus depends crucially on fast decision making. Hence, Tetris provides a clear case of a domain in which connectionist, pattern-completion style reasoning is required for expert performance. If the Dreyfus and Dreyfus model is correct, moreover, such parallel, pattern-completion style reasoning should exhaustively explain expert skill. But interestingly, this does not seem to be so. Instead, expert play looks to depend on a delicate and nonobvious interaction between a fast, pattern-completing module and a set of explicit, higher-level concerns or normative policies. The results are preliminary, and it would be inappropriate to report them in detail. But the key observation is that true Tetris experts report that they rely not solely on a set of fast, adaptive responses produced by, as it were, a trained-up network but also on a set of high-level concerns or policies that they use to monitor the outputs of the skilled network so as to "discover trends or deviations from ... normative policy" (Kirsh and Maglio 1992, 10). Examples of such policies include, "don't cluster in the center, but try to keep the contour flat" and "avoid piece dependencies" (Kirsh and Maglio 1992, 8–9). On the face of it, these are just the kind of rough-and-ready maxims that we might (following Dreyfus and Dreyfus) associate with novice players only. Yet attention to these normative policies seems to mark especially the play of real experts. Still, we must wonder how such policies can help at the level of expert play given the time constraints on responses. There is just no time for reflection on such policies to override online output for a given falling Zoid.

Here Kirsh and Maglio (1992) make a suggestive conjecture. The role of the high-level policies, they suggest, is probably indirect. Instead of using the policy to override the output of a trained-up network, the effect is to alter the focus of attention for subsequent inputs. The idea is that the trained-up network ("reactive module" as they put it) will sometimes make moves that lead to danger situations—situations in which the higher-level policies are not being reflected. The remedy is not to override the reactive module but thereafter to manipulate the inputs it receives so as to present feature vectors that, when processed by the reactive module in the usual way, will yield outputs in line with policy. As they describe it, the normative policies are the business of a distinct planner system that interacts rather indirectly with the online reactive agency: "It is the job of the planner to formulate a specification of concerns. These concerns are translated into directives for changing the focus of attention. Changes in attention in turn affect the feature vector presented to the [reactive agency]" (10). Just how the shift of attention is accomplished is left uncomfortably vague. But they speculate that it could work by "biasing certain board regions" or by "increasing the precision of [certain] values being returned by visual routines" (10).

Despite this vagueness, the general idea is attractive. Effective outputs are always under the control of the trained-up reactive system. But high-level reflection makes a contribution by effectively reconfiguring the input vectors that the reactive agencies receive.

This idea may provide a hint of a solution to the problem of understanding the role of explicitly formulated general commitments (in the form of summary rules or moral maxims) in moral thought.[6] Such commitments—the upshot of individual moral reflection—may help us monitor the outputs of our online, morally reactive agencies. When such outputs depart from those demanded by such policies, we may be led to focus attention on such aspects of input vectors as might help us bring our outputs back into line. Suppose we explicitly commit ourselves to an ideal of acting compassionately in all circumstances. We then see ourselves reacting with anger and frustration at the apparent ingratitude of a sick friend. By spotting the local divergence between our ideal and our current practice, we may be able to bias our own way of taking the person's behavior—in effect, canceling out our representation of those aspects of the behavior rooted in their feelings of pain and impotence. To do so is to allow the natural operation of our on-board reactive agencies to conform more nearly to our guiding policy of compassion. The summary linguistic formulation, on this account, is a rough marker that we use to help monitor the behavior of our trained-up networks.

The moral of the Tetris example, then, is that advanced pattern recognition is really a double skill. In addition to the basic, fluent pattern-recognition-based responses exemplified by a trained connectionist net, the human expert relies on a second skill. This is the ability to spot cases in which these fluent responses are not serving her well. Such recognition (a kind of second-order pattern recognition) is crucial since it can pave the way for remedial action. And it is especially crucial in the moral domain. Here, surely, it is morally incumbent on us not to be hostage to our own fluent daily responses, no matter how well "trained" we are. We must be able to spot situations (for example, dealing with sexual politics in a family setting or interacting with certain religious or political groups) in which these fluent responses are failing to serve us. The effect of formulating some explicit maxims and guidelines provides us with a comparative resource in a sense external to our own online behaviors. This resource is neither binding nor a full expression of our moral knowledge, but it can act as a signpost alerting us to possible problems. The advanced moral agent, like the advanced Tetris player, needs to use every means available to sustain successful performance.

The cases just rehearsed go some way toward correcting the antilinguistic bias discerned in the previous section. Summary linguistic formulations, it seems, are not just tools for the novice. They are tools for the expert too. But the story remains sadly incomplete, for the image of linguistic tools suggests a merely manipulative role. This manipulative role does not, I claim, do justice to the more primary role of linguistic exchange as a medium of genuinely collaborative problem solving. Yet it is under this collaborative aspect (or so I shall argue) that linguistic formulations make their key contribution to moral cognition. It is to this perspective that we now turn.

Language as a Collaborative Medium

Missing from the discussion so far is any proper appreciation of the special role of language and summary moral maxims within a cooperative moral

community. To see this, we can begin by considering the general phenomenon of so-called collaborative learning. The observation here is simply
that a procedure of multiple, cooperative perspective taking often allows
groups of agents to solve problems that would otherwise defeat them. For
example, two children, neither of whom is alone able to come to an
understanding of the Piagetian conservation task (understanding how the
same quantity of liquid can be manifest in very different ways in differently shaped vessels, such as a long, thin glass and a short, fat one) can
often cooperate to solve the problem. The reason is that they "are often
focussing on different aspects of the problem—one saying that the water
in the new beaker is higher and the other noting it is thinner, for example.... These competing perspectives come to light in the interaction, and
in an effort to reach a consensus the children integrate the perspectives,
co-constructing, a new perspective" (Tomasello, Kruger, and Ratner 1993,
501; see also Perret-Clermont and Brossard 1985).

It is the communal effort to achieve consensus that drives the children
to find the solution. Key features of this effort include discussion, joint
planning, critiquing of each other's ideas, and requests for clarification.
Many of these features are transactive in the sense of Kruger (1992). This
means that the thinking and perspective of individual members of a group
are objects of group attention and discussion. Given the crucial role of
such modes of discussion, it is perhaps unsurprising to learn that collaborative learning emerges at about the same developmental moment (age six
or seven) as does so-called second-order mental state talk—talk about other
people's perspectives on your own and others' mental states. Thus, younger children (age three or four) are capable of seeing others as having a
perspective on the world (seeing others as what Tomasello, Kruger, and
Ratner 1993 call mental agents). But it is only the older children who see
others as "reflective agents"—agents whose perspective includes a perspective on the child's own thought and cognition (see Tomasello, Kruger,
and Ratner 1993, 501). Collaborative learning Tomasello, Kruger, and
Ratner argue, requires a participant to recognize others as having ideas
about each other's thoughts and perspectives. It requires participants to
"understand in an integrated fashion the mental perspectives of two or
more reflective agents" (501).

Such a capacity is plausibly viewed as an essential component of advanced moral cognition. Indeed, many moral problems basically consist in
the need to find some practical way of accommodating multiple perspectives, including perspectives on each others' views and interests. Consider
a typical moral issue such as how to accommodate the multiple, and
often competing, perspectives and needs of different religions and racial
groups in a multicultural society. Attempts to find practical solutions to
the kinds of problems thus raised depend crucially on the extent to which

representatives of each group are able to engage in what may be termed multiple nested perspective taking. Consider the case of a conflict within a multicultural educational system.[7] The parents of a Muslim girl requested that she be excused from events involving what (from their perspective) was an unacceptably close physical proximity to boys. The head teacher was inclined to let the child decide. But the likely effect of the child's decision (she did not want to be excluded) would be her total removal from the school. In such a case, the only hope for a practicable solution lies in each party's willingness to try to understand the perspective of the other. It is here, I claim, that the role of linguistic exchange is paramount. The attempts by each party to articulate the basic principles and moral maxims that inform their perspective provide the only real hope of a negotiated solution. Such principles and maxims have their home precisely there: in the attempt to lay out some rough guides and signposts that constrain the space to be explored in the search for a cooperative solution. Of course, such summary rules and principles are themselves negotiable, but they provide the essential starting point of informed moral debate. Their role is to bootstrap us into a kind of simulation of the others' perspectives, which is, as we saw, the essential fodder of genuine collaborative problem-solving activity. No amount of such bootstrapping, of course, can preclude the possibility of genuine conflict between incompatible principles. But it is the exchange of such summary information that helps set the scene for the cooperative attempt to negotiate a practical solution to the problem at hand. Such a solution need not (and generally will not) consist in agreement on any set of general moral rules and principles. Instead, it will be a behavioral option tailored to the specific conflict encountered (see Khin Zaw, unpublished, for just such a defense of "practical reason").

Thus viewed, the rules and maxims articulated along the way are not themselves the determinants of any solution, nor need we pretend that they reveal the rich structure and nuances of the moral visions of those who articulate them. What they do reveal is, at best, an expertise in constructing the kinds of guides and signposts needed to orchestrate a practical solution sensitive to multiple needs and perspectives. This is not, however, to give such formulations a marginal or novice-bound role, nor is it to depict them as solely tools aimed at manipulating all parties into the activation of a common prototype. Rather, it is a matter of negotiating some practical response that accommodates a variety of competing prototypes. (The difference here is perhaps akin to that marked by Habermas's distinction between strategic and communicative action. In strategic action, the goal is to persuade the other, by whatever means, to endorse your viewpoint. In communicative action the goal is to motivate the other to pursue a dialogue by visibly committing oneself to a *negotiated* solution. (Habermas 1990, 58, 59, 134, 145).[8]

The successful use of language as a medium of moral cooperation thus requires, it seems, an additional and special kind of knowing how—one not previously recognized in connectionist theorizing.[9] It concerns knowing how to use language so as to convey to others what they need to know to facilitate mutual perspective taking and collaborative problem solving. The true moral expert is often highly proficient at enabling cooperative moral debate. Moral expertise, *pace* Dreyfus and Dreyfus, cannot (for moral reasons) afford to be mute. This additional know-how, like the other expert skills discussed in the first section, may well itself consist in our commanding a certain kind of well-developed prototype space, but it will be a space that is interestingly second-order in that the prototypes populating it will need to concern the informational needs of other beings: beings who themselves can be assumed to command both a rich space of basic prototypes concerning the physical, social, and moral world and a space of second-order prototypes concerning ways to use language to maximize cooperative potential.

It is perhaps worth remarking, to emphasize the psychological reality of the complex of second-order skills, that high-functioning autistic children (those with basic linguistic skills) show a marked selective deficit in almost all of the areas I have discussed. It is characteristic of such children to show all of the following: no use of self-regulatory speech or inner rehearsal to help them perform a task (compare the Tetris example); very limited grasp of how to use language to achieve communicative goals; complete failure to recognize others as having a perspective on the child's own mental states; and no evidence of collaborative learning, or any other collaborative activity (Frith 1989, 130–145). These children, Frith suggests, are not able to "share with the listener a wider context of interaction in which both are actively involved" or to "gauge the comprehension of listeners" (126). They will use terms that no one else can understand, such as calling seventeen to twenty-five year olds the "student nurses age group" (125), and they "tend not to check whether their speech is actually succeeding and communicating, nor to they show any curiosity as to why a dialogue has broken down" (Baron-Cohen 1993, 512). The linguistic skills of these high-functioning autistics thus leave out all the collaborative dimensions I have been at pains to stress. As a result, Baron-Cohen (1993) raises the possibility that such children are, in a deep sense, acultural: unable to participate in the shared understanding and cooperative action essential to any true cultural group. Oversimplified connectionist models of moral cognition, by marginalizing the collaborative dimensions of moral action, likewise threaten to isolate the moral agent from her proper home, the moral community.

To sum up, it is only in the context of thinking about genuinely collaborative moral activity that the true power and value of principle-invoking moral discourse becomes visible. Summary moral rules and maxims act as flexible and negotiable constraints on collaborative action. Such rules and principles by no means exhaustively reflect our moral knowledge, but they are the expertly constructed guides and signposts that make possible the cooperative exploration of moral space.

Conclusions: Complementary Perspectives on Moral Reason

The kind of exchange between cognitive science and ethics that underlies the present treatment is quite typical. Historically, the bias of computational cognitive science is toward the individual. Ethical theory, by contrast, has concerned itself from the outset with individuals considered as parts of larger social and political wholes. The attempt to formulate a joint image of moral cognition helps correct the historical biases of each tradition. The ethicist is asked to think about the individual mechanisms of moral reason. The cognitive scientist is reminded that moral reason involves crucial collaborative, interpersonal dimensions. Perhaps neither party strictly requires the other to remind it of the neglected dimensions. But in practice, it is often the joint confrontation of the issues that yields progress in the search for an integrated image. In thus striving for a mutually satisfactory vision, we are forced to discover a common vocabulary and to agree on some focal issues, and to the extent that we do so, we prepare the ground for future participants from still other disciplines.

Such long-term benefits aside, the immediate upshot of this discussion is clear: recent connectionist-inspired reflections on moral cognition are probably right in asserting both that moral thinking is fruitfully depicted as a case of prototype-based reasoning and that summary linguistic principles and maxims can therefore provide only an impoverished gloss on the full complexities of our moral understanding. But the associated tendency to marginalize the role of such principles and maxims (to depict them as mere tools for the moral novice; Dreyfus and Dreyfus 1991) is to be resisted. As we saw, such formulations provide powerful tools for the indirect manipulation of moral cognition both in ourselves and others, and, most important, essential signposts and constraints that guide collaborative problem-solving activity. Such collaborative activity is only possible, I argued, courtesy of a special kind of knowing how: a knowing how whose focus is on the informational needs that must be met if others are to participate with us in cooperative problem-solving activity. Such know-how (knowing how to use language to prime the collaborative problem-solving machinery) requires a certain conception of other agents—a conception that recognizes others as already enjoying a particular perspective on the thoughts and viewpoints of their fellows. In the light of all this, we can now see much that is missing from the basic connectionist story. A satisfying story about moral cognition and moral expertise must attend to a variety of thus far neglected, communication-specific, higher-order prototype spaces. To do so will be to recognize that the production and exploitation of summary linguistic rules and principles is not the production and exploitation of mere impefect mirrors of moral knowledge. Rather, it is part and parcel of the very mechanism of moral reason.

Acknowledgments

I extend special thanks to Margaret Walker, Larry May, Marilyn Friedman, Owen Flanagan, Teri Mendelsohn, Peggy DesAutels, the members of the Washington University Ethics Seminar, the Philosophy/Neuroscience/ Psychology work-in-progress group, and the audience at the 1993 Mind and Morals Conference at Washington University in St. Louis.

Notes

1. See Ruth Barcan Marcus, "Moral dilemmas and consistency," in C. W. Gowans (ed.), *Moral Dilemmas* (New York: Oxford University Press, 1987), pp. 188–204. The comments concerning the potential elaboration of the moral code occur on pp. 190–191.
2. "Uneasily," because the typicality findings are not conclusive evidence against a classical view. See Armstrong, Gleitman, and Gleitman (1983) and Osherson and Smith (1981).
3. Not all networks have designated output units, but the basic device of state-space representation characterizes the knowledge acquired even by so-called pattern associator models.
4. "Summary," because extended treatments (such as those of classic literature) may indeed convey detailed information about the structure of moral space. "Summary linguistic expression" refers instead to attempts to distill moral knowledge into short rules and principles.
5. "Postconventional" here refers to stage 6 of Kohlberg's hierarchy of moral development (Kohlberg 1981), a stage at which principles are used to generate decisions.
6. I thank Peggy DesAutels for drawing my attention to the importance of such general normative commitments.
7. This is an actual case, borrowed from Susan Khin Zaw, "Locke and Multiculturalism: Toleration, Relativism and Reason," unpublished manuscript.
8. Habermas often assimilates the idea of strategic action to the idea of the manipulation of others by force or sanctions. Obviously, the idea of manipulation by provision of context-fixing input is importantly different. The question when such provision constitutes genuine manipulation as opposed to collaborative investigation is a delicate and important one. I note in passing that Habermas' emphases also echo those of this treatment in other ways, such as the recognition of the importance of multiple perspective taking (Habermas 1990, 138–146) and the conception of norms as practical, flexible aids rather than rigid defenses (180).
9. This was pointed out to me by Margaret Walker, whose help and comments have improved this chapter in numerous ways.

126   Andy Clark

References

Armstrong, S., Gleitman, L., and Gleitman, H. 1983. "On What Some Concepts Might Not Be." *Cognition* 13:263–308.
Baron-Cohen, S. 1993. "Are Children with Austism Acultural?" *Behavioral and Brain Sciences* 16:512–513.
Bradley, F. H. 1876. "Collision of Duties." In C. W. Gowans, ed., *Moral Dilemmas*. New York: Oxford University Press, 1987.
Churchland, P. M. 1989. *A Neurocomputational Perspective: The Nature of Mind and the Structure of Science*. Cambridge, Mass.: MIT Press.
Churchland, P. M. Forthcoming. "Learning and Conceptual Change: The View from the Neurons." In A. Clark and P. Millican, eds., *Essays in Honour of Alan Turing*. Oxford: Oxford University Press.
Churchland, P. S., and Sejnowski, T. J. 1992. *The Computational Brain*. Cambridge, Mass.: MIT Press.
Clark, A. 1989. *Microcognition: Philosophy, Cognitive Science and Parallel Distributed Processing*. Cambridge, Mass.: MIT Press.
Clark, A. 1993. *Associative Engines: Connectionism, Concepts and Representational Change*. Cambridge, Mass.: MIT Press.
Dreyfus, H., and Dreyfus, S. 1990. "What is Morality? A Phenomenological Account of the Development of Ethical Expertise." In D. Rasmussen, ed., *Universalism vs. Communitarianism: Contemporary Debates in Ethics*. Cambridge, Mass.: MIT Press.
Flanagan, O. 1991. *Varieties of Moral Personality: Ethics and Psychological Realism*. Cambridge, Mass.: Harvard University Press.
Frith, U. 1989. *Autism*. Oxford: Blackwell.
Goldman, A. 1993. "Ethics and Cognitive Science." *Ethics*. 103:337–360.
Habermas, J. 1990. *Moral Consciousness and Communicative Action*, translated by C. Lenhardt and S. Weber Nicholsen. Cambridge, Mass.: MIT Press.
Johnson, M. 1993. *Moral Imagination: Implications of Cognitive Science for Ethics*. Chicago: University of Chicago Press.
Khin Zaw, S. n.d. "Does Practical Philosophy Rest on a Mistake?" Unpublished manuscript.
Kirsh, D., and Maglio, P. 1992. "Reaction and Reflection in Tetris." In J. Hendler, ed., *Artificial Intelligence Planning Systems: Proceedings of the First Annual International Conference AIPS 92*. San Mateo, Calif.: Morgan Kaufman.
Kirsh, D., and Maglio, P. 1994. "On Distinguishing Epistemic from Pragmatic Action." *Cognitive Science* 18:513–549.
Kohlberg, L. 1981. *Essays on Moral Development*. Vol. 1, *The Philosophy of Moral Development*. New York: Harper & Row.
Kruger, A. C. 1992. "The Effect of Peer and Adult-Child Transaction Discussions on Moral Reasoning." *Merill-Palmer Quarterly* 38:191–211.
Land, E. 1977. "The Retinex Theory of Color Vision." *Scientific American* (December): 108–128.
McClelland, J., and Rumelhart, D. 1976. "A Distributed Model of Human Learning and Memory." In J. McClelland, D. Rumelhart, and the PDP Research Group, *Parallel Distributed Processing*. Cambridge, Mass.: MIT Press.
McClelland, J., Rumelhart, D., and the PDP Research Group. 1986. *Parallel Distributed Processing: Explorations in the Microstructure of Cognition*. 2 vols. Cambridge, Mass.: MIT Press.
Marcus, Barcan R. 1987. "Moral Dilemmas and Consistency." In C. W. Gowans, ed., *Moral Dilemmas*. New York: Oxford University Press.
Nagel, T. 1987. "The Fragmentation of Value." In C. W. Gowans, ed., *Moral Dilemmas*. New York: Oxford University Press.

Osherson, D., and Smith, E. 1981. "On the Adequacy of Prototype Theory as a Theory of Concepts." *Cognition* 9:35–38.

Perret-Clermont, A. N., and Brossard, A. 1985. "On the Interdigitation of Social and Cognitive Processes." In R. A. Hinde, A. N. Perret-Clermont, and J. Stevenson-Hinde, eds. *Social Relationship and Cognitive Development.* Clarendon Press, Oxford.

Rosch, E. 1973. "Natural Categories." *Cognitive Psychology* 4:324–350.

Sejnowski, T., and Rosenberg, C. 1987. "Parallel Networks That Learn to Pronounce English Text." *Complex Systems* 1:145–168.

Smith, E., and Medin, D. 1981. *Categories and Concepts.* Cambridge, Mass.: Harvard University Press.

Tomasello, M., Kruger, A., and Ratner, H. 1993. "Cultural Learning." *Behavioral and Brain Sciences* 16:495–552.

Chapter 7
Gestalt Shifts in Moral Perception
Peggy DesAutels

Moral philosophers often assume that there are clear and unambiguous single descriptions of particular moral situations, and thus they view their primary task as that of determining the most moral action to take when in these situations. But surely there is less chance of there being a single and final way to describe a given moral situation than there is of there being a single and final way to organize and describe a visual display. Although we perceive many of our day-to-day moral experiences in an unreflective and even reflexive manner, it is also possible for us to (and we often do) "reperceive" moral situations. On one end of the spectrum, we can slightly adjust our original perceptions by attending to details of moral significance that were at first unnoticed. Or on the other end of the spectrum, we can dramatically shift from our original perceptions to very different moral perspectives or frameworks.

I argue in this chapter that gestalt shifts play a significant role in the mental processes used to determine the moral saliencies of particular situations. I build on the recent debate between Carol Gilligan and Owen Flanagan over the relevance of the gestalt-shift metaphor to the organization and reorganization of our moral perceptions (Gilligan 1987; Flanagan and Jackson 1990; Gilligan and Attanucci 1988; and Flanagan 1991). Throughout the course of this debate, neither of them directly referred to important related issues found in philosophical and psychological discussions of perception. I propose to place this debate within that broader context and argue that a discussion of gestalt shifts in moral perception is directly linked to the more general consideration of how it is that we abstract from and draw meaning out of situations. Connectionist models of cognition, along with research on the role of tasks, metaphors, and analogies in perceptual mental processes, help answer the question, To what degree and under what conditions do we experience gestalt shifts in the organization of our moral perceptions?

Before continuing, it may be helpful to consider a specific example of a gestalt shift in moral perception. Stephen R. Covey provides a dramatic account of "re-seeing" moral saliencies in his best-seller, *The Seven Habits of*

Highly Effective People. Although the description is rather extended, I include most of it. His setting of the scene and his description of how he reframed the situation will be useful for further discussion and analysis:

> I [Covey] remember a mini-paradigm shift I experienced one Sunday morning on a subway in New York. People were sitting quietly— some reading newspapers, some lost in thought, some resting with their eyes closed. It was a calm, peaceful scene.
>
> Then suddenly, a man and his children entered the subway car. The children were so loud and rambunctious that instantly the whole climate changed.
>
> The man sat down next to me and closed his eyes, apparently oblivious to the situation. The children were yelling back and forth, throwing things, even grabbing people's papers. It was very disturbing. And yet, the man sitting next to me did nothing.
>
> It was difficult not to feel irritated. I could not believe that he could be so insensitive as to let his children run wild like that and do nothing about it, taking no responsibility at all. It was easy to see that everyone else on the subway felt irritated, too. So finally, with what I felt was unusual patience and restraint, I turned to him and said, "Sir, your children are really disturbing a lot of people. I wonder if you couldn't control them a little more?"
>
> The man lifted his gaze as if to come to a consciousness of the situation for the first time and said softly, "Oh, you're right. I guess I should do something about it. We just came from the hospital where their mother died about an hour ago. I don't know what to think, and I guess they don't know how to handle it either."
>
> ... Suddenly I *saw* things differently, and because I *saw* differently, I *thought* differently, I *felt* differently, I *behaved* differently [Covey's emphasis]. My irritation vanished. I didn't have to worry about controlling my attitude or my behavior; my heart was filled with the man's pain. Feelings of sympathy and compassion flowed freely....
> Everything changed in an instant. (Covey 1989, 30–31)

Note that Covey refers to this experience as a "mini-paradigm shift." Any discussion of gestalt shifts in moral perception will certainly overlap with recent discussions of "paradigm shifts" in scientific theories.[1] However, I refer to a shift in the organizing of a particular moral situation as a gestalt shift to emphasize that not all shifts are as incommensurable as Kuhn's paradigm shifts.

Some might argue that shifts in moral perception are few and far between, but I argue that gestalt shifts, are common and unavoidable, and they play a significant role in the mental processes used to perceive particular moral situations. More specifically, I argue that gestalt shifts range from shifts between "unmergeable" or "rival" *details of* a perceived situation to shifts between *entire organizing* perspectives of a situation; and these shifts play a significant role in the mental processes used to determine the moral saliencies of particular situations. I differentiate between our more unreflective day-to-day moral perceptions (which incorporate what I term *framework* shifts as a result of switching tasks) and our more deliberative moral "perceptions" (which include both framework shifts and what I term *component* shifts).

Gilligan: Shifts between Care and Justice Perspectives

Carol Gilligan, a moral psychologist, has focused much of her work on differences in moral reasoning between male and female research participants. In her early studies, subjects were presented with short descriptions of hypothetical dilemmas and then asked to reason morally about these dilemmas. More recently, Gilligan has asked her subjects (both male and female) to supply and describe their own examples of moral dilemmas. She switched to soliciting real-life moral dilemmas in order to discover which experiences the subjects themselves most viewed as moral dilemmas. As a result of this work, Gilligan offers the following psychological moral theory:

> 1) Concerns about justice and care are *both* represented in people's thinking about real-life moral dilemmas, but people tend to focus on one set of concerns and minimally represent the other; and 2) there is an association between moral orientation and gender such that both men and women use both orientations, but Care Focus dilemmas are more likely to be presented by women and Justice Focus dilemmas by men. (Gilligan and Attanucci 1988, 88)

Gilligan views unfairness-fairness and inequality-equality as those saliencies most associated with the justice perspective, and attachment-detachment and responsibility-irresponsibility as those saliencies most associated with the care perspective (Gilligan 1987, 20). In the passage above, Gilligan acknowledges that the same person may be able to organize moral experiences from either the justice or the care perspective, but she argues that one perspective predominates. When we do view a particular situation using both perspectives, Gilligan maintains that we shift between perspectives rather than organize that particular experience by at once combining justice and care concerns.

I will not discuss the degree to which justice, or care, or both predominates in men's and women's overall moral orientations. Rather, I will describe Gilligan's views on shifts. Gilligan compares the shift between justice and care perspectives to visual gestalt shifts that take place when viewing ambiguous figures. She writes:

Like the figure-ground shift in ambiguous figure perception, justice and care as moral perspectives are not opposites or mirror-images of one another, with justice uncaring and care unjust. Instead, these perspectives denote different ways of organizing the basic elements of moral judgment: self, others, and the relationship between them. (Gilligan 1987, 22)

Although admitting that more research needs to be done on "whether, for example, people alternate perspectives, like seeing the rabbit and the duck in the rabbit-duck figure, or integrate the two perspectives in a way that resolves or sustains ambiguity" (Gilligan 1987, 26), she does argue that we vacillate between rather than integrate the justice and care perspectives. She refers to the "focus phenomenon," whereby subjects lose sight of one group of potential saliencies (justice saliencies) and focus instead on another group (care saliencies) in moral perception.

Gilligan's evidence for the contention that we shift between perspectives that are "not readily integratable" is garnered from research subjects' verbal descriptions of moral situations. Subjects were deemed to have switched perspectives when the terms they used in one analysis of the moral dilemma did not contain the terms of another analysis of the same dilemma. She refers to the doctoral research of Kay Johnston in which children and teenagers were presented with a moral dilemma in fable form. According to Johnston, about half the children "spontaneously switched moral orientations" (they switched terminology) when asked if there was another way to "solve the problem" (Gilligan 1987, 26–27).

It is worth noting that Gilligan leaves room for the possibility that there are more than two moral orientations when she refers to "*at least* two moral orientations" (Gilligan 1987, 26, my emphasis). She does not, however, conjecture on the nature or composition of any additional moral perspectives beyond care and justice.

Flanagan: Integrated Moral Perceptions

Owen Flanagan takes a more interdisciplinary approach to moral cognition and brings recent advances in cognitive science, moral theory, and psychology to bear on the issues surrounding moral perception. In his discussion of Gilligan's two perspectives, Flanagan primarily wishes to take issue with the degree to which persons have a dominant orientation, but he also objects to certain aspects of Gilligan's use of the gestalt-shift metaphor. He describes three ways in which the metaphor is helpful and even illuminating but follows this discussion with three ways in which it is "unhelpful and misleading" (Flanagan 1991, 214). His primary objections to the metaphor are: (1) "Not all moral issues are so open to alternative construals";

(2) "Moral consideration, unlike visual perception, takes place over time and can involve weighing as much, and as messy, information as we like"; and (3) "The metaphor calls attention to the gross features of moral perception [but] much of what is most interesting and individually distinctive about moral personality lies in the small details of what is noticed, deliberated about, and acted on" (Flanagan 1991, 214–217).

In the following sections, I bring recent developments in cognitive science to this discussion of the role of gestalt shifts in organizing moral experience.

Gestalt Shifts

I do not limit the definition of a gestalt shift to examples of completely polar and/or incommensurable organizations. Rather I include any two mental organizations that cannot be merged into one (or are incompatible in some way), no matter how much these two organizations may have in common. Another way to put this is to use the visual focus metaphor. When one set of saliencies comes into focus, most members of that set (but not necessarily all) go out of focus when the competing set of saliencies comes into focus. The duck-rabbit image (figure 7.1) is a more extreme example of a visual image that can result in our experiencing two distinct and incompatible perceptions. When we see the duck, we do not see the rabbit; and after switching to seeing the rabbit, we do not see the duck. However, images with figure-ground ambiguities can also result in the experience of a gestalt shift (figure 7.2). In figure-ground shifts, there is

Figure 7.1
Duck-rabbit ambiguous figure. From *Mind Sights* by Roger N. Shepard. Copyright © 1990 by Roger N. Shepard. Reprinted with permission of W. H. Freeman and Company.

Figure 7.2
Figure-ground ambiguity. From *Mind Sights* by Roger N. Shepard. Copyright © 1990 by Roger N. Shepard. Reprinted with permission of W. H. Freeman and Company.

more of an emphasis on refocusing, where background becomes foreground and foreground becomes background, rather than an emphasis on incompatible and completely distinct figure interpretations.[2] Gilligan and I define gestalt shifts as incorporating both incompatible figures and figure-ground ambiguities, whereas Flanagan emphasizes more the view that gestalt shifts (and thus gestalt shift metaphors) necessarily include a more radical figure incommensurability.

The experience of a gestalt shift in moral perception may or may not be provoked by an external stimulus. Just as it can be mentioned to someone "stuck" seeing only a rabbit that the image can also be seen as a duck (and the "bill" of the duck can even be directly pointed to), so can someone supply new information that encourages the reperceiving of a moral situation. Covey's shift, for example, was preceded by new information from the man in the subway. But one could also imagine Covey's reperceiving the situation with no external stimulus. For instance, he could have suddenly noticed that these children were behaving similarly to children he had observed previously at a parent's funeral service.

When a shift in moral perception is preceded by an external stimulus, there is a sense in which the situation itself has changed. I wish to emphasize, however, that gestalt shifts can occur even when the "input" to a perception changes to some degree. Gestalt shifts occur in such cases when the new piece of perceptual input is not simply added on or incorporated into the original perception; rather, at least some of the original perceptual input array is mentally reorganized.

It is also important to this discussion to distinguish between a framework gestalt shift and a component gestalt shift (hereafter referred to as framework shift and component shift). What I term a framework shift involves (1) a mental switch from one way of organizing an entire experience to a different way of organizing that experience and (2) some sense in which the two ways of organizing the experience are incompatible—in other words, the two overall organizations cannot be merged into a single overall organization. The duck-rabbit switch would be a framework shift if the duck-rabbit image filled our entire viewing screen as it were—if the duck-rabbit image comprised our perceptual experience.

On the other hand, what I term a component shift involves (1) a mental shift from one way of organizing a detail (component) of an experience to a different way of organizing that detail, and (2) some sense in which the two ways of organizing that detail of experience are incompatible. A visual example of a component shift is the shift that would occur when viewing the duck-rabbit image as part of a larger scene. For instance, we may see a person wearing a t-shirt with many patterns on it, one of which is the duck-rabbit pattern. The overall organization of the scene remains the same (a person wearing a t-shirt with patterns on it), but a detail in the scene shifts. For the purposes of this discussion, then, a gestalt shift occurs whenever the perceiver shifts between the deployment of one already-existing organizing mental structure to the deployment of another already-existing organizing structure in the perceiving of at least some (if not all) of a particular situation or experience.

Moral perceptions can involve either framework or component shifts. For example, we could continue to frame a particular woman's abortion dilemma around the status of the fetus (rather than reorganizing the frame around the relationship of that woman to the fetus) but shift between viewing the fetus first as an unborn child and then as a simple growth. Such a shift would be a component shift in moral perception. With these clarifications in mind, let us move on to consider when and under what conditions gestalt shifts in moral perception are most likely to take place.

Tasks and Gestalt Shifts in Moral Perception

As we go about our daily lives, it only makes sense that our current tasks would heavily influence which of all possible perceptual organizations (possible for us with the learning history that each of us has) is actually brought online. If there is something we are trying to accomplish, we will notice and abstract out those aspects of our experience that most help us to achieve our objective. Our days are filled with such tasks as fulfilling job requirements, running errands, getting x done, solving problem y, and so on.

Neither Gilligan nor Flanagan directly addresses the relationship between task switching and gestalt shifts in moral perception. In this section, I argue that in our more unreflective day-to-day moral perceptions, an important relationship exists between switching tasks and gestalt shifts in moral perception. The obvious result of acknowledging this relationship is the additional acknowledgment that there are many more gestalt shifts in our day-to-day moral perceptions than either Gilligan or Flanagan postulates and many of these shifts are between perceptual organizations other than the justice-care organizations.

An inherent difficulty in analyzing the relationship between perceptual organizations and tasks is that of determining the best level of description for particular tasks. A task can be described at any number of levels of generality or complexity. To keep this discussion of tasks simple, I describe someone as engaged in a particular task (mental or otherwise) if it makes sense that the person would describe herself as currently engaged in that task if asked (for example, "What are you doing right now?" "I'm trying to figure out what's fair.")

In *Varieties of Moral Personality*, Flanagan relates the perceptual process of abstracting out certain features of a situation to a person's current task. He specifies two main types of abstraction: feature detection (or classificatory abstraction) and task-guided abstraction. These two main types of abstraction are not mutually exclusive mental processes but interact in complex ways. He defines feature detection as involving "the cognitive isolation or recognition of just those properties (sometimes called essential properties) which warrant classifying some token as a member of a type or kind" (83–84). He defines task-guided abstraction as the deployment of rationalized procedures that "warrant paying differential attention, and giving differential treatment, to various features of an object, event, or situation" in order to complete a given task successfully (85). All kinds of abstraction involve highlighting or separating out the features of some thing or event that bring it under the correct cognitive description relative to its actual nature and the aims of the person doing the abstracting. Task-guided abstraction goes further and involves deployment of rationalized procedures deemed appropriate to the successful completion of the task at hand. These procedures warrant paying differential attention, and giving differential treatment, to various features of an object, event, or situation. My point is this: to the degree that our perceptions are task guided, we will shift our perceptions and the organized saliencies of our perceptions when we shift tasks. Thus, switches to significantly different tasks will often involve what I consider to be gestalt shifts in perception.

Of course, not all task switches precipitate shifts in moral perception. Since neither mowing my lawn nor raking the cuttings is likely to include the perception of any moral saliencies, it is highly unlikely that switching from mowing to raking will result in a gestalt shift in moral perception. So it is important at this point to examine what kinds of day-to-day tasks and task switches most affect moral perceptions. Clearly, switching between ostensibly "moral" tasks will be most likely to affect moral perceptions. For example, switching from the task of determining what is most fair to that of determining what is most caring may well result in our reorganizing our experience.[3] Most of us are not engaged in self-described moral tasks much of the time. We may live by such high-level moral goals as that of responding morally when it is called for, but the specific task of responding morally may be initiated only intermittently in our day-to-day experience.

This is not to say that we do not make moral judgments and determine moral saliencies unless engaged in a self-described moral task. In fact, any task that includes interactions with others will often incorporate the making of moral judgments, assessments, or both. The most obvious day-to-day task likely to involve moral perceptions is our "job."[4] For instance, the moral qualities of those who most determine our job success will often be the qualities of most salience to us throughout our workday.

To illustrate, while engaged in solving a problem at work with a colleague, I am most likely to organize my perceptual experience around the obstacles and means to solving this work-related problem. If my colleague is preventing the project from proceeding, I view this colleague as "difficult" (at best), and all of her personality faults become highly salient in my experience. However, I may also have a high-level goal of treating others with compassion when they are "in need." Even in the middle of an intense, work-related discussion, I may notice something out of sorts with my colleague, "interrupt" my foreground work-related task, and initiate my "treat others with compassion" moral task.[5] As a result, I switch to reperceiving the saliencies in my colleague's behavior most important to my responding with compassion. In this case, the shift in my moral perceptions is tied to my switching from an ostensibly "nonmoral" task (which nonetheless incorporates moral perceptions) to a "moral" task.

I can also switch, for example, from a more general moral task of attempting to determine the most fair thing to do in a situation to the more specific moral task of seeing things from another person's point of view. This switch from a general to a more specific moral task also results in reorganizing the moral saliencies of the situation at hand.

Even Gilligan's examples of shifting from the justice to the care perspective can be viewed as switching between the moral task of determining what is most fair and the moral task of determining what is most caring. Certainly some may not wish to describe this change in orientation as so closely intertwined with a task switch. There are surely other ways to describe the reorientation, but viewing the shift as tied to task switching accomplishes two objectives: (1) I emphasize that the perceiver's goals and activities have much more relevance to the perceived moral saliencies of a situation than most others have presumed, and (2) I deemphasize the role of inherent aspects of a situation itself and emphasize instead how much of moral perception is directed by the perceiver.[6] In our day-to-day lives, different and competing saliencies occur primarily because we switch to different tasks with their accompanying different objectives.

Covey's mini-paradigm shift described at the beginning of this chapter could also be viewed as tied to his switching from the moral task of determining what would bring most peace and harmony to the passengers on the subway to the moral task of determining what would bring most comfort and compassion to the family whose mother had just died. There may have been others on the subway car who were entirely focused on their own tasks of whatever sort and thus failed to notice anything other than the fact that there were some loud but ignorable children in the "background." Interestingly, it could even be argued that Covey experienced two gestalt shifts. The first was the shift in perception tied to his switching from a nonmoral reading task to the moral task of helping to bring about peace and quiet on the car, and the second was the shift tied to his switching to a distinctly different moral task of responding compassionately to a grieving family.

Past Experience, Analogies, Metaphors, and Connectionist Prototypes

Although switching tasks plays a significant role in day-to-day shifts in moral perception, it can also be argued that switching analogies, metaphors, and even concepts while engaged in an overtly moral task (such as moral deliberation or moral reflection) also results in gestalt shifts in moral perception. Chalmers, French, and Hofstadter (1992) describe "high-level perceptual processing" (involving the drawing of meaning out of situations) as consisting of a complex interaction between the process of making analogies ("mapping one's situation to another") and the process of perceiving the situation ("filtering and organizing [data] in various ways to provide an appropriate representation for a given context") (192–195). Although they do not directly address either moral perception or gestalt shifts in perception, they note that there can be "rival analogies." For example, we can shift between viewing Saddam Hussein as being like

Hitler and viewing him as being like Robin Hood, a "generous figure redistributing the wealth of the Kuwaitis to the rest of the Arab population" (199). It is doubtful that Americans would make such a shift spontaneously. We would have very little incentive to shift analogies in order to view Hussein as Robin Hood unless we made it our moral task, for example, to understand how his own people view him.

Connectionist models of the mind and perception also imply that we can experience gestalt-like shifts when our task becomes that of better understanding a situation. Churchland's analysis of "conceptual change versus conceptual redeployment" is of direct relevance to a discussion of gestalt shifts. He describes "conceptual redeployment" as "a process in which a conceptual framework that is already fully developed, and in regular use in some other domain of experience or comprehension, comes to be used for the first time in a new domain" (Churchland 1989, 237). Churchland gives the example of Huygens's applying his already well-developed "wave" conceptual framework to his understanding of light for the first time.

What is not clear in Churchland's account is how often and under what circumstances we "conceptually redeploy" in our deliberations and everyday lives. Many day-to-day gestalt shifts in perception are precipitated by the conscious switching of tasks, but there is also no doubt that sudden and unexplainable framework shifts periodically occur in our lives. If we conceive of conceptual redeployments merely as spontaneous framework shifts, redeployments seem few and far between. On the other hand, conscious switching to a rival analogy or a different metaphor should also count as conceptual redeployments. If so, we "conceptually redeploy" quite regularly. Both framework and component shifts in moral perception can be viewed in connectionist terms as switching between the activation of one moral prototype to the activation of a different (and in some way incompatible) prototype in order to make sense of particular situations.

Churchland does briefly examine the relationship of his connectionist model of the mind to moral theory. His description of what occurs in moral argument closely parallels my emphasis on gestalt shifts in moral perception. For Churchland, moral argument takes place when situations are ambiguous and "consists in trying to reduce the exaggerated salience of certain features of the situation, and to enhance the salience of certain others, in order to change which prototype gets activated" (Churchland 1989, 300).

Another interesting consequence of viewing knowledge as prototypical (in a connectionist sense of the word) is that we come to see knowledge as much more context dependent. In other words, the representations that lade our perceptions are context-dependent representations. Andy Clark expands on this notion when he writes,

Fodor-style classicists were seen to picture the mind as manipulating context-free symbolic structures in a straight-forwardly compositional manner.

Connectionists, not having context-free analogues to conceptual-level items available to them, have to make do with a much more slippery and hard-to-control kind of "compositionality" which consists in the mixing together of context-dependent representations. (Clark 1993, 25)

To illustrate what is meant by a context-dependent representation, Clark refers to Smolensky's "infamous coffee case." We experience coffee in a variety of contexts (liquid coffee in a cup, ground coffee in a can, and so on), and Clark argues that as a result, we do not have a single representation for coffee; rather, we have many different "coffee" representations. In connectionist terminology, a "coffee" representation comprises a set of activation patterns in the hidden units; these patterns vary because the contexts varied in which the net was "trained up" on coffee (Clark 1993, 24).

In the light of my definition of a gestalt shift, switching from one context-dependent representation of "coffee" to another context-dependent representation of "coffee" is tantamount to a gestalt shift. After all, it is a shift between two incompatible and "unmergeable" mental structures. We cannot unify our various coffee activation patterns into one all-purpose pattern but must switch between differing context-dependent mental structures depending on the context in which we are deploying the coffee concept. Nonetheless, it is important to reemphasize that there are degrees to which we shift between representations as we attempt to draw meaning out of a situation. These range from framework shifts (where a new and complex "wave" prototype is brought to bear on Huygens' perception of "light") to component shifts (where in a deliberation over a situation involving coffee, the perceiver can bring one of several coffee activation patterns to bear on the situation). It is interesting to note that even "coffee" can be tied to a moral perception. We could, for example, be engaged in the moral task of determining whether our being addicted to coffee is harmful to others. As we deliberate over this "addiction situation," we may shift between the activation of various context-dependent "coffee" representations.

The pervasiveness of gestalt shifts in connectionist models of the mind can also be illuminated by considering recent work of Mark Rollins on the plasticity of perception. He emphasizes plasticity in use or the strategy ladenness of perceptual knowledge and points out that much of our perceptual experience incorporates strategies having to do with which of our already-existing concepts to deploy in a given situation. He writes that "plasticity-in-use is important because it can produce a change in the effectiveness of content even if not in content itself" (Rollins 1994, 42). Once again, the point can be made that whenever we restrategize content use, whenever we redeploy concepts, we have shifted in a gestalt-like manner in our perceiving of a situation.

Flanagan Revisited

With the knowledge that switching between tasks, analogies, metaphors, and even contextualized concepts will result in gestalt shifts in our perceptions, it is worth reexamining Flanagan's objections to the gestalt shift metaphor. His first concern is that "not all moral issues are open to alternative construals" (Flanagan 1991, 214). But I have stressed that the experiencing of gestalt shifts in moral perception is not determined primarily by the nature of the situation. Rather, in our more unreflective day-to-day moral perceptions, shifts are often tied to task switches. In our more deliberative moral "perceptions," we can attempt a different perspective on any moral issue by attempting to see the situation "through someone else's eyes" or by applying a different analogy or metaphor.

Flanagan's second objection to the gestalt-shift metaphor is that "moral consideration, unlike visual perception, takes place over time and can involve weighing as much, and as messy, information as we like" (215). However, as I have pointed out, even moral deliberation or consideration over time of a particular situation may incorporate mental structures that conflict and cannot be merged into single "messy" mental entities. We cannot weigh and then merge as much, and as messy, information as we like when the information comes in context-dependent pieces.

Flanagan stresses that perceptions involving cognition and deliberation result in "all-things-considered judgments" (communication with Flanagan 1994). I agree that a deliberative moral perceiver has an ability to weigh alternatives and then arrive at an "all-things-considered" perspective on a situation. My point here is that even this consideration process will often involve shifting between gestalt-like, context-dependent mental structures. In other words, deliberative moral perceptions often involve selecting from various incompatible organizing structures rather than constructing a single "best" perspective using fine-grained, context-free mental elements. Flanagan appears to assume that moral deliberation consists in the manipulation of and recomposition of context-free representations, but connectionist models of mind give us no such fine-grained, context-free mental elements with which to "build" moral perceptions.

Flanagan's final concern is that the gestalt shift metaphor "calls attention to the gross features of moral perception [but that] much of what is most interesting and individually distinctive about moral personality lies in the small details of what is noticed, deliberated about, and acted on" (Flanagan 1991, 217). While what I term framework shifts stress the larger-grained overall organization of moral perception, there remain finely detailed and richly articulated saliencies within each framework. Shifts in and of themselves make no difference to the content of what we shift from or to. For example, when we shift to a rival analogy, we still have all the richness and detail of that analogy cognitively available. I have also argued throughout this chapter that not all shifts are framework shifts; many gestalt shifts are, rather, component shifts in how we view particular details. In many moral situations involving several actors, for example, we may not reframe the entire situation but shift significantly in our view of one of the "players."

Conclusion

This discussion has linked moral perception with gestalt shifts by showing how we bring already-existing mental structures to bear on situations. Altering our perceptions by shifting between already-existing mental structures does not by any means comprise all of perceptual cognition— moral or otherwise. After all, much of learning involves the altering of previously existing mental structures and the creating of new mental structures. And much of moral thought involves reasoning—using the perceived saliencies of a situation as "input" to higher-level, traditionally rational mental processes. But it should also be clear that much of perceptual cognition itself—the drawing of meaning from situations—incorporates the application of various of our already-in-place concepts and conceptual organizations.

Acknowledgments

I am especially grateful to Andy Clark, Owen Flanagan, Marilyn Friedman, Larry May, and Mark Rollins for their helpful comments on earlier drafts of this chapter.

Notes

1. For more on the related philosophy of science discussion of gestalt shifts, see Wright (1992).
2. Marilyn Frye (1983) proposes this type of figure-ground refocusing, for example, in situations where men have been viewed as the main actors and women as "stagehands" in the background; we can refocus these scenes so that the men recede and the women come into focus.
3. Switching from determining what is most fair to determining what is most caring may or may not result in our reorganizing our experience. If, for example, what is most fair also turns out to be what is most caring, it may be that no reorganization occurs. For more on how the concepts of justice and care can overlap, see Friedman (1987).

4. Sara Ruddick (1989) has similarly argued that the day-to-day work of mothers determines the kind of moral thinking they do.
5. My use of computer-task terminology here is not meant to convey a flippant attitude toward moral cognition. I take morality and the living of a moral life very seriously.
6. Since discussions of moral perception are often linked to discussions of moral realism, I should mention that I am neither a moral realist nor do I think that cognitive scientists and/or ethical theorists provide any compelling reasons for being so.

References

Chalmers, David, Robert M. French, and Douglas Hofstadter. 1992. High-level Perception, Representation, and Analogy: A Critique of Artificial Intelligence Methodology. *Journal of Expert Theory and Artificial Intelligence* 4:185–211.
Churchland, Paul. 1989. *A Neurocomputational Perspective: The Nature of Mind and the Structure of Science.* Cambridge, Mass.: MIT Press.
Clark, Andy. 1993. *Associative Engines.* Cambridge, Mass.: MIT Press.
Covey, Stephen R. 1989. *The Seven Habits of Highly Effective People: Restoring the Character Ethic.* New York: Simon & Schuster.
Flanagan, Owen. 1991. *Varieties of Moral Personality: Ethics and Psychological Realism.* Cambridge, Mass.: Harvard University Press.
Flanagan, Owen, and Kathryn Jackson. 1990. Justice, Care, and Gender. In Cass Sunstein, *Feminism and Political Theory.* Chicago: University of Chicago Press.
Friedman, Marilyn. 1987. Beyond Caring: The De-Moralization of Gender. In Marsha Hanen and Kai Nelson, eds., *Science, Morality, and Feminist Theory.* Calgary: University of Calgary Press.
Frye, Marilyn. 1983. *The Politics of Reality: Essays in Feminist Theory.* Trumansburg, N.Y.: Crossing Press.
Gilligan, Carol. 1987. Moral Orientation and Moral Development. In Eva Feder Kittay and Diana T. Meyers, eds., *Women and Moral Theory.* Savage, Md.: Rowman & Littlefield.
Gilligan, Carol, and Jane Attanucci. 1988. Two Moral Orientations. In C. Gilligan, J. V. Ward, J. M. Taylor, and B. Bardridge, *Mapping the Moral Domain.* Cambridge, Mass.: Harvard University Press.
Gilligan, C., J. V. Ward, J. M. Taylor, and B. Bardige. 1988. *Mapping the Moral Domain.* Cambridge, Mass.: Harvard University Press.
Hanen, Marsha, and Kai Nielson, eds. 1987. *Science, Morality, and Feminist Theory.* Calgary, Alberta, Canada: University of Calgary Press.
Kittay, Eva Feder, and Diana T. Meyers, eds. 1987. *Women and Moral Theory.* Savage, Md.: Rowman & Littlefield.
Rollins, Mark. 1994. The Encoding Approach to Perceptual Change. *Philosophy of Science* 61:39–54.
Ruddick, Sara. 1989. *Maternal Thinking: Toward a Politics of Peace.* New York: Ballantine Books.
Shepard, Roger N. 1990. *Mind Sights: Original Visual Illusions, Ambiguities, and Other Anomalies, with a Commentary on the Play of Mind in Perception and Art.* New York: W. H. Freeman and Company.
Sunstein, Cass R. 1990. *Feminism and Political Theory.* Chicago: University of Chicago Press.
Wright, Edmond. 1992. Discussion: Gestalt Switching: Hanson, Aronson, and Harre. *Philosophy of Science* 59:480–486.

Chapter 8
Pushmi-pullyu Representations
Ruth Garrett Millikan

I Introduction

A list of groceries, Professor Anscombe once suggested, might be used as a shopping list, telling what to buy, or it might be used as an inventory list, telling what has been bought (Anscombe 1957). If used as a shopping list, the world is supposed to conform to the representation: if the list does not match what is in the grocery bag, it is what is in the bag that is at fault. But if used as an inventory list, the representation is supposed to conform to the world: if the list does not match what is in the bag, it is the list that is at fault. The first kind of representation, where the world is supposed to conform to the list, can be called directive; it represents or directs what is to be done. The second, where the list is supposed to conform to the world, can be called descriptive; it represents or describes what is the case. I wish to propose that there exist representations that face both these ways at once. With apologies to Dr. Doolittle,[1] I call them pushmi-pullyu representations (PPRs) (figure 8.1).

PPRs have both a descriptive and a directive function, yet they are not equivalent to the mere conjunction of a pure descriptive representation and a pure directive one but are more primitive than either. Purely descriptive and purely directive representations are forms requiring a more sophisticated cognitive apparatus to employ them than is necessary for these primitives. Purely descriptive representations must be combined with directive representations through a process of practical inference in order to be used by the cognitive systems. Similarly, purely directive representations must be combined with descriptive ones. The employment of PPRs is a much simpler affair.

This chapter originally appeared in *Philosophical Perspectives*, Volume 9, edited by James Tomberlin (Ridgeview Publishing, 1995). It is reprinted, with minor revisions, by permission of the author and publisher.

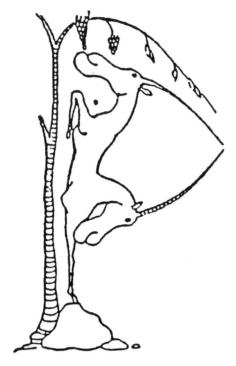

Figure 8.1
Pushmi-pullyu.

Perhaps the most obvious PPRs are simple signals to conspecifics used by various animals, such as bird songs, rabbit thumps, and bee dances. But PPRs also appear in human language and probably in human thought. Illustrations in language are, "No, Johnny, we don't eat peas with our fingers" and "The meeting is adjourned," as said by the chair of the meeting. Human intentions are probably an example of PPRs in thought, serving at once to direct action and to describe ones future so that one can plan around it. Our inner representations by which we understand the social roles that we play as we play them are probably also PPRs. The natural way that we fall into doing "what one does," "what women do," "what teachers do," and so forth suggests this. I suspect that these primitive ways of thinking are an essential glue helping to hold human societies together.

I view PPRs from within a general theory of representations developed in other places (Millikan 1984, 1993). I will start by sketching just a little of that theory (though not nearly enough to defend it).

II The Background Theory of Representation

Brentano took the essence of intentionality to be the capacity of the mind to "intend" the nonexistent. In recent years it has become generally accepted that he was right in this sense: the core of any theory of representation must contain an explanation of how misrepresentation can occur. I have argued that misrepresentation is best understood by embedding the theory of intentionality within a theory of function that allows us to understand more generally what malfunction is. For this, we use a generalization of the biological notion of function as, putting things very crudely, the survival value of a reproduced type of entity. I call functions of this generalized kind "proper functions" (Millikan 1984, 1993).

Think of the proper function of a type as what it has been doing to account for its continued reproduction. The possibility of misrepresentation is derived from the possibility that a token may fail to perform a function that has been accounting for continued reproduction of its type. In some cases, failure to function properly may be statistically even more common than proper functioning. For example, many biological mechanisms fail much of the time to have those occasional propitious effects that have nonetheless accounted for their proliferation in the species. Thus, the eyeblink reflex, exhibited when any object moves too close to the eye, may be triggered uselessly many times for every time it actually prevents foreign matter from entering the eye. In order to proliferate, often a type needs to perform properly only in some critical but small proportion of cases, a proportion that varies widely depending on other factors that enter into the mechanisms of evolution by natural selection.

We can apply the notion of proper function directly to natural languages. Whole sentences are not usually reproduced, but phonemes and words are reproduced, and the syntactic forms in which they are placed are reproduced. They are copied from one generation of speakers to the next and reproduced by the same speaker on various occasions. If some theory of universal grammar is correct, then certain very general grammatical features are also reproduced via the genes. But more immediately, it is clear that speakers of a given language reproduce words patterned into certain concrete syntactic forms rather than certain other forms because the effects these patterns have had upon hearers, in some critical proportion of cases, are effects the speakers wish to reproduce. These effects are described with reference to the semantic and conventional pragmatic rules of the language.

The proper effects upon hearers of language forms—the proper reactions of hearers to these forms—are also reproduced. This is not necessarily because hearers directly copy one another's reactions, though this may sometimes happen. These reactions are reproduced because hearers, at least often enough, benefit from reacting properly to the sentences they hear. By moving from sentences into belief or into action in conformity with the rules of the language, often enough hearers gain useful knowledge or their actions find a reward; hence, they reproduce conformity to these rules. Thus, each generation of hearers learns to accord in its reactions with the expectations of the previous generation of speakers, and each generation of speakers learns to expect those same reactions from a previous generation of hearers—with sufficient frequency, that is, to keep the language forms in circulation. Similarly, the proper reaction of a bird to the song of a conspecific, though not for most species an imitated response, is also reproduced—in this case, genetically reproduced. In both cases, the proliferation of representations and the proliferation of proper reactions to them are each dependent on and tailored to the other.

The proper function (or functions) of an expression in a public language may contrast with the function that a speaker intends for it on a given occasion. I have shown how the speaker's intention in use can lend tokens of a language device additional proper functions, as in the case of Gricean implicature (Millikan 1984, chap. 3). These additional functions are not functions of the public language forms, however. Indeed, the two layers of function, public and private, can sometimes conflict, as in the case of purposeful lying and certain parasitic uses of language. We will not need to consider this sort of complexity here, but it is well to warn in advance that if there are pushmi-pullyu forms existing in the public language, as I will argue there are, these should not be confused with the well-known phenomenon of Gricean implicature. Public pushmi-pullyu forms are double-aspected on the first layer of function—the public language layer. It is not merely the user's intention that produces the pushmi-pullyu effect.

To understand what inner representations are (percepts, thoughts), we apply the notion of a proper function not directly to the representations themselves but to the mechanisms whose function it is to produce and to use inner representations. When functioning properly, inner-representation-producing mechanisms produce representations in response to, and appropriate to, situations in which the individual organism finds itself. In humans, these mechanisms are exceedingly complex, including mechanisms of belief and desire formation and also mechanisms of concept formation, inference, decision making, and action. When the entire system functions properly, the belief-forming mechanisms produce true beliefs and the desire-forming mechanisms produce desires whose fulfillment would benefit the organism. But it is possible also to sharpen the notion "proper function" in such a way that inner representations themselves are seen to have proper functions, as follows.[2]

Many biological mechanisms are designed to produce alterations in the organism in response to some aspect of the environment, so as to adapt or match the animal to that aspect, hence to serve some further function within that environment. A simple example is the mechanism in the skin of a chameleon that rearranges its pigment so that its color will match that of its environment. This mechanism then serves the further function of concealing the chameleon from predators. Each particular coloring of the chameleon produced is naturally said to have a function too—the same function: concealing the chameleon from predators. Similarly, any state constituting a stage in a (proper-)functional process, when the shape of the process and hence of the state is determined by input from and as a function of certain features of the environment, can be viewed as itself having proper functions. The proper functions of this state are to help in the production of various further stages in the process to which this state will give rise if the whole system continues to function properly. In this way, specific inner states, even quite unique inner states, unique because they are induced by unique organism-environment relations, may have proper functions. In this way, even such representations as someone's desire to climb Mount Everest backward, or someone's belief that persimmons cure mumps, can be considered to have proper functions.

But I have not yet said what any of these proper functions are—what it is that representations, either inner or outer, properly do.

III Descriptive and Directive Representations

Elsewhere I have defended a proposal to explain how the content of a representation—its satisfaction condition—is derived. The explanation takes time, and it is different for descriptive and for directive representations. Here I will assume the notion of content, explaining only what I take the difference between descriptives and directives to be.

A representation is directive when it has a proper function to guide the mechanisms that use it so that they produce its satisfaction condition. Like a blueprint, it shows what is to be done. Desires are directive representations. To see how this might be so, remember that the proper function of an item can be a function that it is unlikely to perform. Perhaps the sad fact is that an overwhelming majority of our desires never become satisfied. Many (for instance, the desire to square the circle) may even be incapable of becoming satisfied or (for instance, the desire that it rain tomorrow) may be incapable of being satisfied by normal operation of those mechanisms that are designed to help fulfill desires. This has no bearing on the claim that desires are satisfied when things proceed "properly," that is, in the ideal sort of way that accounted for the survival and proliferation of those integrated systems whose job it is to make and use desires. Surely the job of these systems is, first, to produce desires that would benefit one if fulfilled and then, second, given certain additional propitious inner and outer circumstances, to be moved by them to their fulfillment.

Sentences in the imperative mood are, paradigmatically, directive representations. They proliferate in the language community primarily insofar as they (often enough) help to effect the fulfillment of their satisfaction conditions.

Unlike directive representations, what makes a representation descriptive is not its function. Rather, the descriptive representation's truth condition is a condition to which it adapts its interpreters or users in the service of their proper functions (Millikan 1984, chap. 6; 1993, chaps. 4–6). It is a condition that must hold if the interpreter's (proper) way of being guided by the representation is to effect fulfillment of the interpreter's functions in accord with design. For example, beliefs are descriptive representations. If my belief that there is an umbrella in the hall closet is to help guide my decision-making and action-guiding apparatus (the belief's "interpreter") such that it serves its function of helping to fulfill my desires—for example, my desire to keep off the rain—then there needs to be an umbrella in the hall closet. If it is to help me make a correct inference concerning, for example, whether Susan returned my umbrella, one that yields truth not by accident but in accordance with the good design of my cognitive systems, still there needs to be an umbrella in the hall closet, and so forth.

Typical sentences in the declarative syntactic pattern have a descriptive function. Their function is to produce hearer beliefs or, more precisely, to produce true hearer beliefs. For it is only when a certain proportion of what hearers are told is true, hence is interpreted by them into true beliefs, that they are encouraged to continue to conform to certain rules of interpretation from English (from the public language) into beliefs. One could not learn to understand a language if too large a proportion of the sentences one heard in it were false. Because its function is to produce true beliefs in hearers, the declarative syntactic pattern is descriptive. Roughly speaking, sentences in this pattern affect their interpreters in a proper way only under the conditions that are their truth conditions.

Earlier I remarked that I was not pausing to explain how the satisfaction conditions of representations are determined, but it will be important to grasp this much. On the analysis I have given, the satisfaction condition of a representation is derived relative to its function. The content of the representation turns out to be an abstraction from a fuller affair intrinsically involving an embedding mood or propositional attitude. Put simply, there is no such thing as content without mood or attitude; content is an aspect of attitude. A corollary, as we will soon see, is that it is possible for the very same representation to carry at once two different contents, one relative to each of two different attitudes or moods that simultaneously embed it.

IV Pushmi-Pullyu Representations

Consider first a very primitive representation: the food call of a hen to its brood. A proper function of this call is to cause the chicks to come to the place where the food is and so to nourish them. Assume, what is reasonable, that this is the only proper effect that the call has on chicks, the only effect the call has been selected for. Then the call is directive, saying something like, "Come here now and eat!" But it is also a condition for proper performance of the call that there be food when the hen calls, so the call is also descriptive, saying something like, "Here's food now." (Note that the descriptive and the directive contents of this representation are different.) Assume further, what is again reasonable, that the effect of the call on the chicks is not filtered through an all-purpose cognitive mechanism that operates by first forming a purely descriptive representation (a belief that there is food over there), then retrieving a relevant directive one (the desire to eat), then performing a practical inference, and finally, acting on the conclusion. Rather, the call connects directly with action. *Its function is to mediate the production of a certain kind of behavior such that it varies as a direct function of a certain variation in the environment, thus directly translating the shape of the environment into the shape of a certain kind of conforming action*: where the hen finds food, there the chick will go. The call is a PP representation.

Other examples of primitive PPRs (probably) are other bird calls, danger signals used by the various species, the various predator calls used by chickens and vervet monkeys, and the dance of the honey bee. For example, the bee dance tells at once where the nectar is and where to go. Functioning properly, it produces variation in behavior as a direct function of variation in the environment. Actually, there is evidence that the bee has a map in its head of its environment and that the dance induces it, first, to mark the nectar location on this map (Gallistel 1990).[3] Still, assuming that the only use the bee ever makes of a mark for nectar on its inner map is flying to the marked position to collect the nectar, then the nectar on the bees' inner map is itself a PPR. And it seems reasonable to count a representation whose only immediate proper function is to produce an inner PPR as itself a PPR.

James J. Gibson did not advocate speculating about inner representations. Yet his notion that in perception we perceive certain affordances (opportunities for action) suggests that perceptual representations are PPRs. Think of perceptual representations simply as states of the organism that vary directly according to certain variations in the distal environment. The perceived layout of one's distal environment is, first, a representation of how things out there are arranged—a descriptive representation. It is also a representation of possible ways of moving within that environment:

ways of passing through, ways of climbing up, paths to walk on, graspable things, angles from which to grasp them, and so forth. Variations in the layout correspond to variations in possible projects and in the paths of motion needed to achieve them. The representation of a possibility for action is a directive representation. This is because it actually serves a proper function only if and when it is acted upon. There is no reason to represent what can be done unless this sometimes affects its being done. Compare desires, which serve a function only insofar as they occasionally help to produce their own fulfillment. In the case of perceived affordances, action toward their fulfillment is, of course, directly guided by the percept, variations in the environment, hence in the percept, translating directly into variations in the perceiver's movement.

There are cells in the inferior premotor cortex of monkeys (informally, "monkey see—monkey do cells") that fire differentially according to the immediate ends (such as grasping small pieces of food with the fingers) of the manual manipulations the monkeys are about to execute and that also fire when the monkeys see other monkeys perform these same manipulations for the same ends (Rizzolatti et al. 1988). Imitative behaviors in children show up extremely early. One infant was observed in the laboratory to imitate facial expressions (opening the mouth, sticking out the tongue, etc.) at the age of forty-two minutes (Meltzoff and Moore 1983). We might speculate, on analogy with the monkeys, that these primitive mechanisms of imitation in children employ PPRs, which picture what the other is doing at the same time that they serve to direct what the self is to do. Compatibly, Jeannerod (1994) cites evidence that imagining oneself performing certain movements and actually performing them involve, in part, the same dedicated area of the brain, hence, that picturing what one might do and intending to do it may be two sides of numerically the same representing coin. Indeed, one of his suggestions is that imagining an action without at the same time performing it is accomplished by inhibition of normal connections to motor pathways.

It is important to see that PPRs are not merely conjunctions of a descriptive plus a directive representation. They are more primitive than either purely directive or purely descriptive representations. Representations that tell only what the case is have no ultimate utility unless they combine with representations of goals, and, of course, representations that tell what to do have no utility unless they can combine with representations of facts. It follows that a capacity to make mediate inferences, at least practical mediate inferences, must already be in place if an animal is to use purely descriptive or purely directive representations. The ability to store away information for which one has no immediate use (pure description), and to represent goals one does not yet know how to act on (pure direction), is surely more advanced than the ability to use simple kinds of PP representations.

V PP Representations in Human Thought

Organisms often evolve in complexity by modifying less differentiated multipurpose structures into more differentiated dedicated ones. Thus, we would expect beliefs (dedicated to facts) and desires (dedicated to goals) to be a later evolutionary achievement than inner PPRs. On the other hand, if there are purposes that could be served as well or better by PPRs than by more differentiated representations, our first hypothesis should be that that is how these purposes still are served. I think that there are some such purposes.

One obvious hypothesis is that human intentions are PPRs. If intentions are inner representations, surely they are at least directive ones. They perform their proper functions when they issue in the intended actions. But it is also a common and plausible assumption that a person cannot sincerely intend to do a thing without believing she will do it. If one starts in a rather traditional way, assuming that there are only two basic sorts of cognitive representations—purely descriptive ones (beliefs) and purely directive ones (desires)—then whether "intending" must involve harboring a descriptive as well as a directive thought may appear to be a matter of "analysis of the concept 'intention.'" But if intentions are PPRs, then the dual nature of intentions is no conceptual truth but a biological or neurologic truth. And there is reason to suppose they might be PPRs.

Suppose that my brain already harbors, for purposes of guiding my action, a representation of what I am definitely going to do. And suppose there is need to take this settled future into account when making further decisions about what else I can compatibly do. It would surely waste space and introduce unnecessary mechanisms for evolution to duplicate the representation I already have. Better just to use it over again as a descriptive representation as well. Notice, however, that this kind of PPR differs from the kinds I have previously discussed in this way. Rather than functioning as do, say, perceptual PPRs, which map variations in the organism's world directly into (possible) actions, it maps variations in goals directly onto the represented future world. It differs also in that the contents of the directive and descriptive aspects of the representation are not different but coincide.

A second kind of PPRs that may be fundamental in human thought are primitive representations of social norms and roles. I suggest not that this is the only way humans can cognize these norms and roles but that it may be the primary functional way, and that this way of thinking may serve as an original and primary social adhesive. There are good reasons for thinking that humans and other social animals have designed into them mechanisms leading to the coordination of behaviors.[4] Coordinated behaviors are ones that benefit each individual involved in the coordination given that the others also play their assigned roles. Some of the principles governing evolution of such behaviors are now well known. Allan Gibbard suggests that "systems of normative control in human beings ... are adapted to achieve interpersonal coordination" (1990, 64). His ultimate aim is to cast light on the origins and function of the language of ethics and the thought it communicates, including especially the function of normative discussion in originating what he calls "normative governance." My project here is less ambitious. My speculations concern only a mechanism for stabilizing and spreading coordinative behaviors that are already in place and for coordinating expectations.

I have in mind two basic sorts of social coordinations. The first might be called *common* norms. They apply equally to all members of a given society: we drive on the right; we speak at meetings only when duly recognized; we wait in orderly queues; we are quiet at concerts; we honor our contracts; we see to our families first, then to our relatives, then to our friends; and so forth. The second might be called *role* norms. These apply to a person only insofar as he or she is filling a certain role: children obey adults while adults direct children; the chair of the meeting calls it to order and so forth, but does not introduce or speak to motions or vote, while the members do introduce motions, speak to them, and vote; pupils raise hands to be called on while teachers speak freely; guests and hosts behave in assigned ways; and so forth.

The examples of norms just mentioned undoubtedly all have coordinating functions. On the other hand, the distinction between common norms and role norms could be applied as well to norms lacking coordinating functions. For example, not eating peas with one's fingers and not picking one's nose in public may be noncoordinating common norms, whereas wearing a skirt if you are female and trousers if you are male may always have been a noncoordinating role norm. A mechanism whose biological function is to transmit coordinating norms might well have as a mostly benign side effect the transmission of a good number of noncoordinating norms as well. Thus, humans tend to be creatures of convention, exhibiting many patterns of both solo and interactive behavior that are handed on quite blindly, seeming to serve no purpose at all. It may be that our propensity to play games is another side effect of a mentality built to effect coordinations.

The mechanism for stabilizing coordinative patterns of behavior that I propose is simple. It is the capacity and disposition to understand social norms in a way that is undifferentiated between descriptive and directive. What one does (or what *das Mann* does—remember Heidegger?), what a woman does, what a teacher does, how one behaves when one is married or when one is chair of the meeting, these are grasped by means of thoughts, PPRs, that simultaneously describe and prescribe. In the primitive mind, these PPRs describe and prescribe what is understood to be The

Moral Order: an order taken as totally objective, noninstrumental (absolute), and real, but understood at the same time as stringently prescribed. (In primitive thought, self and others have Sartrian essences with a vengeance.) But it may also be that without the general disposition to think in this way during much of the unreflective parts of our lives, the social fabric would be weakened beyond repair. Yes, I am seriously proposing this as a possible neural mechanism, although supplemented, of course, with more sophisticated mechanisms by which we moderns may also dissect the relevant norms to reveal two faces.

VI PP Representations in Human Language

If human thought contains PPRs, arguably human languages should contain them as well. Just as they contain forms, on the one hand, whose function is to implant ideas about what is the case and, on the other, to implant ideas about what to do, one would expect them to contain forms whose function is to implant mental PPRs in hearers. For example, if the inner vehicles of our intentions and our unreflective graspings of social norms are PPRs, it is reasonable that there should be linguistic forms to correspond. There do not seem, however, to be any dedicated syntactic arrangements to do these jobs. And indeed, granted PPRs really exist in thought, this lack of a form of expression dedicated entirely to PPRs may do much to explain their near invisibility. PPRs, I believe, are imparted by use of the declarative syntactic pattern. At least this is true for English. That is, the declarative pattern has more than one function. Sometimes it is descriptive, and sometimes it expresses a PP mood.

In the PP mood we say, for example, to children, "We don't eat peas with our fingers" and "Married people only make love with each other." The job of this mood is to describe and to prescribe, producing at the same time both true expectations and coincident behaviors, the one as a direct function of the other. Notice that the mechanism here is not that of Gricean implicature. Both functions are explicit or literal; the mood is proliferated precisely because it serves both functions at once; both functions are fully conventional.

Strict orders are standardly delivered in the English declarative pattern, which then functions directively: "You will report to the CO at 6 A.M. sharp"; "You will not leave the house today, Johnny, until your room is clean." This use, I suggest, is more than just directive; it is not just another imperative form.[5] Its function is to impart an *intention* to a hearer and to impart it directly, without mediation through any decision-making process, for example, without involving first a desire and a practical inference. This is the PP mood, undifferentiated between directive and descriptive, serving to impart PPRs.

The PP mood may also be used to impart intentions to a group, thus serving a coordinating function. "The meeting will now come to order," when it functions properly, imparts to each member of the group both intentions concerning their own behavior and expectations concerning the behavior of others. The PP mood statement in the university catalog, "Professors will hold office hours every day during registration week," informs both students and faculty and imparts intentions to faculty.

I would like finally to introduce two more kinds of PPRs that I believe occur in public language but that will require somewhat more lengthy discussions. The first is a variety of explicitly performative sentence. The second are terms expressing "thick concepts."

VII Performatives

Many performative utterances, I believe, are in the PP mood.[6] When viewed incorrectly as simple present-tense descriptives, they are puzzling in that they seem to create facts *ex nihil*, making something to be the case simply by saying that it is the case. Suppose that the chair of the meeting says, "The meeting is adjourned," but nobody pays any attention and three more motions are debated. There does seem to be a sense in which these motions will have been debated not just after the meeting was *called* adjourned but after it *was* adjourned, that is, a sense in which "the meeting is adjourned" is self-guaranteeing. Similarly, once the chair has said, "The meeting will now come to order," the meeting has been brought to order even if everyone keeps talking loudly. And if the right person says to the right couple at the right time, "I now pronounce you man and wife," then these two are man and wife, even if they do not act that way and even if nobody, including the legal authorities, is disposed to treat them that way. On the other hand, if nobody pays any attention when the chair says, "The meeting is adjourned" or "The meeting will now come to order," there is also a tug that says these sentences somehow were not true. Let me try to explain why this tension occurs.

Many conventionally or legally molded patterns of activity allow or require the making of, as I will call them, conventional or legal "moves," under designated circumstances, by participants' playing designated roles —for example, performing a marriage ceremony, making a move in a game, appointing someone to a position, making a motion in accordance with parliamentary procedure, or legally closing a road. These are acts that, once performed, place restrictions on what may or must follow after if things are to accord with convention or law. Moves of this sort may or may not be made with the aid of articulate language forms. For example, in some contexts a bid may be made by saying, "I bid," but in others, merely by raising a finger, a vote may be made by raising a hand, and so forth.

Most such moves are themselves in essence PPRs.[7] Their proper function is to channel behaviors that follow so that they take certain forms and not others (directive) and to coordinate expectations accordingly (descriptive). Functioning properly in the usual way, they produce inner PPRs in the various participants, which guide them in coordinating their expectations and actions.

The implications of such moves for ensuing behavior are often quite complex, involving complicated mandates or the limiting of options for a variety of affected people playing a variety of roles. For this reason, these implications often could not easily be spelled out in an explicit formula, for example, by saying something simple like, "The meeting will now come to order." Imagine the minister's trying to fill in all the blanks in this formula: "You, Jane, and you, John, will now ... and the guardians of law will now ... and your parents and friends will now ..." and so forth, spelling out all that marriage entails. However, there usually are names for conventional or legal moves, or for the situations constituted by their having been performed: "bidding six diamonds," "checking the queen," "being married," "being chairman," "appointing a chairman," "voting for a candidate," "a road being legally closed," "making a motion," and so forth. These names are not names for the "shapes" of the moves—they are not names like "raising one's finger" or "raising one's hand." Rather, they are names or descriptions that classify moves according to their conventional outcomes—by what sorts of things follow after them in accordance with convention or law.

Now there is an obvious and very general metaconvention by which one may use the name of any move in order to make that move, for, assuming that the conventional "shape" that the move has is merely conventional, as it generally is, any other shape might be substituted for this shape so long as everyone understands what move is being made. Saying that I perform the move will generally be enough to perform it, then, granted that the vehicle that usually performs it is arbitrary, and granted that I am the fellow whose move it is to make. For example, even, "I move my queen to Q5" may be quite enough to perform that move in chess, especially if circumstances are such as to make an actual board move awkward or impossible (such as chess by mail or while also trying to get dinner).

In English, present-tense declaratives are used in making such moves: "I bid three diamonds," "I pronounce you man and wife," "I move that the meeting be adjourned," "The meeting is adjourned," "This road is legally closed," and so forth.[8] This use of the declarative pattern proliferates because it serves the function of producing the conventional outcomes of the moves named, channeling ensuing activities (a directive function) and coordinating relevant expectations (a descriptive function) at the same time. The proper function of the declarative in each of these cases is thus exactly the same as the function of the move named or described; indeed, the move IS the utterance, in the right circumstances, of the sentence naming it, just as raising one's hand in the right circumstances is voting. It is a matter of convention that one can vote by raising one's hand, and it is a matter of (very sensible) convention that one can make almost any conventional move by embedding its name in this way in the declarative pattern. These embeddings are called "performative formulas." Such performative formulas are PPRs. They produce conventional outcomes and coordinate expectations accordingly, standardly by producing relevant inner PPRs in participants.

But there is another detail. One can, of course, make a conventional move with a performative formula that names that move only granted that one is the right person in the right circumstances to make that move. I cannot, for example, adjourn a meeting by saying, "The meeting is adjourned," unless I am chair and we are in the right kind of a meeting and the right time has come. Another function of the performative formula—another thing it does when functioning properly—is to produce true beliefs that the named conventional move is in fact being performed. In this respect, the performative is just (is also) an ordinary descriptive. It is true if it is in fact being uttered by an appropriate person in appropriate circumstances, that is, if the move it claims is performed is in fact being performed. Nor does this hang on whether it in fact effects its conventional outcome.

Suppose that I put up a sign on my road saying, "This road is legally closed." If I have reproduced this token of the declarative pattern on the model merely of descriptive sentences I have previously heard rather than copying my use from the cultural pattern of performative uses, then it is not, of course, a sentence in the PP mood at all. It is merely descriptive, and, assuming the town has not in fact closed my road, it is plain false. If I have reproduced it instead on the model of PP sentences I have previously heard, perhaps supposing myself to have the legal right to close my own road in this way, then it is in the PP mood. Still, just as if it were a simple descriptive, it is, minimally, false. It is not true that my road is legally closed.

Integrating this now with the previous theme, besides this last sort of truth condition, a performative sentence of this kind also has a directive satisfaction condition and a second truth condition as well. It directs that a certain conventional outcome should ensue, and it induces expectations to accord with this. Returning to, "The meeting is adjourned," one of its truth conditions is fulfilled by the fact that it is the chair who says it and at an appropriate time. Whether or not three more motions are debated afterward, once the chair has said these words, the meeting is in fact adjourned.

There is a second truth condition, however: the proceedings should then actually draw to a close. Only if this happens will the coordinated expectations that it is the proper job of the performative sentence to induce be true expectations. Last, like any PP mood sentence whose job is to impart intentions, it has the meeting's actually drawing to a close as a directive satisfaction condition as well.

VIII Thick Concepts

I wish now to speculate about one last example of PPRs in public language: sentences that contain words expressing "thick concepts." Thick concepts are concepts such as *rude, glorious,* or *graceful,* which seem to describe a thing and to prescribe an attitude toward it at the same time. Recently there have been absorbing and complex discussions about these concepts that I cannot enter into here. But I would like make a suggestion that might be fruitful to explore.

I hesitate to take any strong position on thick concepts in part because I am not sure, within the framework I have been using, how to understand the nature of those states or dispositions that ordinary speech calls "attitudes." (Philosophers' "propositional attitudes" of belief and desire are not "attitudes" in ordinary speech at all.) It is not likely, for example, that an adequate description of the attitude involved in thinking something rude could be given by reference only to the notions "directive" and "descriptive." But let us suppose this much: if the function of a term in the language is in part to induce an attitude, then this part of its function is not entirely descriptive. That is, attitudes are other than or more than mere (descriptive) beliefs. If I can make plausible, then, that one function of a term expressing a thick concept is to produce an attitude but that these terms are also descriptive, I shall have shown at least that these terms are close kin to PP representations.

Suppose that certain (perhaps highly disjunctive) configurations of primary qualities tend to produce in us certain attitudes toward their bearers when perceived or contemplated. This might be due to native dispositions or the influence of culture. Secondary qualities are traditionally thought of as powers to produce sensations, but powers to produce attitudes are perhaps similar enough to be called secondary qualities too—"attitudinal secondary qualities." Secondary qualities are such only relative to a kind of perceiver or, in this case, a kind of reactor. Nonetheless, relative to a certain group or class of reactors, attitudinal secondary qualities are objective properties.

My suggestion is that words expressing thick concepts may, first, describe attitudinal secondary qualities relative either to our species as a whole or to the culture of speaker and hearer. That is, declarative sentences using these words attributively will not serve their proper functions in a normal way unless the objects of attribution do indeed have certain attitudinal secondary qualities relative to a community encompassing both speaker and hearer. But, second, this is because their proper function is a directive one: to produce in the hearer the relevant attitudes. Their function is to cause the hearer to take these attitudes toward these items. That is, these words continue to produce in hearers these attitudes toward designated objects only because the attitudes induced turn out, in a large enough proportion of cases, to be independently "true" by the hearers' own lights. The objects described are such as actually to produce, on direct inspection or given a more detailed description, those attitudes. Words expressing thick concepts thus face two ways at once, describing their subjects and at the same time inducing attitudes toward these subjects. Indeed, perhaps the inner representations induced by these words themselves face two ways at once. Perhaps they are inner PPRs.

Notes

1. And Hugh Lofting: *The Story of Doctor Dolittle; Being the History of His Peculiar Life at Home and Astonishing Adventures in Foreign Parts.*
2. See, especially, Millikan (1984) chap. 2; 1993, chap. 11; 1994.
3. If the dance says there is nectar at a location where the bee's experience has previously shown there is a large body of water, the bee will be unmoved by the dance (Gallistel 1990).
4. A good informal summary of these reasons is given in Gibbard (1990).
5. This is a correction of the remarks on army orders in Millikan (1984, chap. 3).
6. The performative sentences most obviously referred to in the following discussion are those that perform what Strawson (1964) termed "essentially conventional illocutionary acts" or, more accurately I believe, "K-II illocutionary acts" as these are described in Millikan (forthcoming), sect. VI. The relation of these performatives to those that perform explicit "K-I illocutionary acts" is interesting but too involved to pursue in a general essay of this sort.
7. I say "most" because some conventional patterns have no functions at all. See Millikan (forthcoming).
8. Some have argued that performatives function by conversational implicature. However, explicit performatives like "I order you to," "I promise that," and so forth are lacking in some languages, "particularly those with a more developed system of sentence types or those spoken in societies that seem to have less cultural need for formulaic discourse of the kind represented by performative sentences" (Sadock 1988, 186).

References

Anscombe, G. E. M. 1957. *Intention.* Ithaca: Cornell University Press.
Gallistel, C. R. 1990. *The Organization of Learning.* Cambridge, Mass.: Bradford Books/MIT Press.
Gibbard, A. 1990. *Wise Choices, Apt Feelings.* Cambridge, Mass.: Harvard University Press.
Jeannerod, M. 1994. "The Representing Grain: Neural Correlates of Motor Intention and Imagery." *Behavioral and Brain Sciences* 17:187–245.

Meltzoff, N., and Moore, K. 1983. "Newborn Infants Imitate Adult Facial Gestures." *Child Development* 54:702–709.

Millikan, R. G. 1984. *Language Thought and Other Biological Categories*. Cambridge, Mass.: MIT Press.

Millikan, R. G. 1993. *White Queen Psychology and Other Essays for Alice*. Cambridge, Mass.: MIT Press.

Millikan, R. G. 1994. "A Bet with Peacocke." In C. Macdonald and G. Macdonald, *Philosophy of Psychology: Debates on Psychological Explanation*. Oxford: Oxford University Press.

Millikan, R. G. Forthcoming. "Proper Function and Convention in Speech Acts." In L. E. Hahn, ed., *The Library of Living Philosophers*. La Salle, Ill.: Open Court.

Rizzolatti, G., Carmada, R., Fogassi, L., Gentilucci, M., Luppino, G., and Matelli, M. 1988. "Functional Organization at Area 6 in the Macaque Monkey. II. Area F5 and the Control of Distal Movements." *Experimental Brain Research* 71:491–507.

Sadock, J. M. 1988. "Speech Act Distinctions in Grammar." In *Linguistics: The Cambridge Survey*. Vol. 2. Cambridge: Cambridge University Press.

Strawson, P. F. 1964. "Intention and Convention in Speech Acts." *The Philosophical Review* 73(4):439–60.

Part III
Moral Emotions

Chapter 9
Sympathy, Simulation, and the Impartial Spectator
Robert M. Gordon

Hume observed that our minds are mirrors to one another, reflecting one another's passions, sentiments, and opinions. This "sympathy," or "propensity we have to sympathize with others, to ... receive by communication [the] inclinations and sentiments [of others], however different from, or even contrary to, our own," he held to be the chief source of moral distinctions (Hume 1739/1978). Hume gave an account of how this mirroring of minds works. After a brief presentation of the account, I will show how it needs to be updated and corrected in the light of recent empirical research. Then I will give some reasons to think that the mirroring of minds is more pervasive than even Hume had thought: that mirroring is an essential part of the way in which we think about other minds. Finally, I will make some remarks about the relevance of mirroring to ethics.

Facial Empathy

In the *Treatise of Human Nature* Hume explained how sympathy works: "When I see the effects of passion in the voice and gesture of any person, my mind immediately passes from these effects to their causes, and forms such a lively idea of the passion as is presently converted into the passion itself" (Hume 1739/1978). That is, first we move inferentially from the effects of an emotion to the idea of the emotion that caused them; then, somehow, the same emotion gets synthesized in us. As Hume expresses it, "the ideas of the affections of others are converted into the very impressions they represent, and ... the passions arise in conformity to the images we form of them" (Hume 1739/1978).

But this account seems implausible. It makes cognition and inference essential links in the communication of emotion, and it requires a mechanism for subsequently converting ideas back into emotions. One problem

This chapter originally appeared in *Ethics* 105 (July 1995) 727–742. It is reprinted with the permission of the University of Chicago Press.

is that the hypothesized dependence on cognition and inference would make it hard to explain the well-known fact that even infants pick up the emotions of others by contagion. Darwin describes the reaction of his six-month-old infant son when the child's nurse pretended to cry: "I saw that his face instantly assumed a melancholic expression, with the corners of his mouth strongly depressed; now this child could rarely have seen any other child crying, and I should doubt whether at so early an age he could have reasoned on the subject. Therefore it seems to me that an innate feeling must have told him that the pretended crying of his nurse expressed grief; and this through sympathy excited grief in him" (Darwin 1896). To account for the infant's expression, Darwin posits "an innate feeling" that gives the child an immediate, noninferential cognition of the nurse's emotion. This seems an improvement over Hume's account. But Darwin's account, like Hume's, makes the communication of emotion hinge on cognition: something has to "tell" the child what emotion the crying expressed. And then there must be a mechanism for converting this cognition into the emotion of which it is a cognition.

Recent experimental work suggests that at least some part of the communication of inclinations and sentiments Hume spoke of, and probably all empathy observed in infants, travels by way of a purely noncognitive channel—one that requires no recognition of the other's mental state and no mechanism for converting cognitions into what they are cognitions of. It has been observed that even newborn infants have a strong tendency to mimic the behavior of other human beings, and there is reason to think that such behavioral mimicry is a channel through which high-level central states such as emotions may be transferred from one individual to another.

First, it is well known that very young infants cry in response to the cry of another infant. This appears to be a prewired response calibrated to the specific sonic parameters of the cries of human infants. In the case of Darwin's son, one possibility is that he suppressed the vocalization (something within the capacity of an infant of six months) but failed to suppress the accompanying facial expression—the downturned corners of the mouth, and so forth. (Facial expressions are far less likely to be suppressed than cries, for we normally hear our own cries but do not typically see our own facial expressions.) Alternatively, Darwin's child may simply have been mimicking the nurse's facial expression as she pretended to cry. Even newborn infants have been found to have a remarkable penchant for motor mimicry, particularly a capacity to copy the facial expressions of others, whether these are expressive of emotions or not. The response appears to originate in a prewired mapping: visually perceived faces translate directly into motor output. That is why infants do not need visual feedback that would enable them to compare their own motor output with what they observe on the other's face.[1] In adults, the response is typically subliminal, probably because overt mimicry tends to get inhibited, but the effect is measurable nevertheless.

There is good evidence that six-month-old infants pick up from the crying of others not just the exterior display of grief or distress but the genuine article. And there is at least one plausible explanation how behavioral mimicry could be a channel for the transfer of central states such as emotions. It is known that motor activity, especially the movement of facial muscles, can drive the emotions. Even when we voluntarily contract specific facial muscles or produce vowel sounds that force the contraction of these muscles, we tend to bring on the emotion that characteristically produces a corresponding facial expression. The behavior brings about changes in reported mood and emotion and induces physiological changes similar to those that typically accompany the emotion. This would explain how, by replicating the facial expressions of others, we would tend to "catch" the emotions of others (Meltzoff and Gopnik 1993), a process I call facial empathy.

Higher Forms of Empathy

Thus far I have described a sympathetic response of a very primitive sort. For one thing, the response is not about anything in particular. One may acquire second-hand distress, but it is not yet distress about something. If a mother is smiling because she is pleased about her promotion, the sight of her smile may produce a smile in her infant, which may in turn produce pleasure in the infant. But it will not be pleasure about anything in particular, certainly not about her mother's promotion.[2]

In more typical cases, where the source of pleasure is something in the immediate environment, the problem of transmitting the intentionality of the emotion is typically solved by another form of low-level mimicry: mimicry of direction of gaze. Like many other animals, human beings automatically turn their eyes in the direction in which another individual is gazing, especially if the other is looking with fixed gaze and exhibiting startle or some other reactive emotion, or in other ways shows attentiveness and interest. We triangulate to the first salient target. Normally the response emerges in the child's first year.

The gaze-tracking response transfers one's attention from the other's facial expression to the "cause" or "object" of the emotion or attitude it expresses. In the process called social referencing, the sequence is typically reversed. One first sees an object or scene that is unfamiliar, puzzling, or possibly threatening and then looks toward another person—in the case of children, typically a parent or caretaker—whose eyes are directed toward the same object or scene. Then by contagion, one picks up the other person's response to the object or scene. In either case, whether one assumes the other's expression and then tracks the gaze to the source or starts with some object or scene and then assumes the expression of someone looking at it, facial empathy together with gaze mimicry can be a powerful socializing mechanism. It allows young children to pick up adult ways of reacting to various features of the environment. The child can learn, for example, to discriminate what is distressing from what is not, what is pleasing from what is not, and what is startling from what is not.

From Contagion to Prediction

Even with a device for transmitting the intentionality of an emotion, the process I have described still amounts to nothing more than emotional contagion. You can catch an emotion, just as you can catch a cold, without knowing whom you caught it from. Facial empathy will be of further help once we have the capacity to refer our second-hand response to the particular other individual whose visage is inducing the response—in other words, to mark the response as part of our representation of that individual. Once we have the capacity to index our response in this fashion and to integrate it with other information about the same individual, facial empathy can assist us in interpreting, predicting, and explaining behavior. We can also, at least to a degree, segregate these copycat responses from ones that arise in the normal way from our own perceptions and memories. Specifically, we can allow the copycat responses to run "off-line," as it were, so that the emotion's typical effects on one's behavior are diminished, if not entirely eliminated.

Just as facial empathy uses our own face as a mirror of the other's face, and uses our own emotions, running off-line, as a kind of simulation of the other's emotions, so too we often use our own "original" responses to the environment in order to get information that we were unable to read off the other's face. Suppose we see someone look startled, or frightened, or obviously pleased about something but cannot easily identify the source of the emotion. Following the other's gaze, we find several objects or environmental features, any one or more of which might be the source. We need to pick out the right feature or features. Or suppose the person has already turned away from the source of the emotion. In such cases, we look around for a plausible target; that is, we look for something startling. Or if the other is frightened, we look for something that is frightening. If pleased, we look for something pleasing. That is, we look for something that is startling, or frightening, or pleasing *to us*. To do this, we engage our own system for generating emotions out of our perceptions. We are also prepared to make adjustments as necessary. For example, I see a competitor for an award and find her looking elated. On the wall nearby there is posted a list of award winners. My own name on the list would indeed be pleasing, but I automatically shift to viewing the list through her eyes. We do this sort of thing so routinely that we are not aware of doing it, and we fail to appreciate the sophistication of the maneuver we are engaging in.

Now suppose we wish to predict how the other will act and want to fit the information gained by facial empathy together with information concerning the other's situation, social status, cultural background, and, most important, the other's actions past and present. How do we do that? The answer depends on how we think such information is represented and put to work in predicting behavior. In broad strokes, I will depict two very different accounts: the "theory" theory and the simulation theory.

People are remarkably competent at predicting the behavior of others and attuning their own behavior accordingly. "I have the impression," Jerry Fodor remarks, "that we manage ... better [with one another] than we cope with less complex machines" (Fodor 1987). This is so because, in his view, we have a rather good theory of how this particular machine—the human behavior control center, the brain—works. This is the so-called belief-desire theory, alias commonsense or folk psychology—a theory that is supposedly possessed by people of all cultures and virtually all levels of intelligence. It is said to posit unobservable mental states such as beliefs, desires, intentions, and feelings, linked to each other and to observable behaviors by "lawlike" principles. These principles are applied to observable situations by way of logical inferences that generate predictions and explanations of behavior. This "theory" view has been the prevailing view in philosophy and the cognitive sciences, even if many would disagree with Fodor's understanding of the specific nature of belief-desire theory and his claim that its universality indicates that it is innate, like a Chomskian universal grammar.

Hot and Cold Methodologies

Fodor is one of the few "theory" theorists to offer a detailed account of the way this alleged folk psychological theory might be put to work. In the first chapter of *Psychosemantics* (1987), he discusses a scene from *A Midsummer Night's Dream*. Lysander, cast under a magic spell, abandons Hermia while she is asleep in the forest. Hermia just cannot believe Lysander would leave her voluntarily, so she accuses his rival Demetrius of murdering Lysander. Fodor suggests that Hermia's accusation is a product of "implicit, nondemonstrative, theoretical inference." Demetrius, being a rival, would want Lysander out of the way. And Demetrius believes: Lysander will not be out of the way unless he is dead, and I, Demetrius, can bring about Lysander's death. Hermia plugs these ascriptions into the law: If one wants that P, believes that not-P unless Q, and also believes one can bring it about that Q, then (ceteris paribus) one tries to bring it about that Q. Hence Hermia says to Demetrius, "It cannot be but thou hast murder'd him."

Hermia comes to think that Demetrius had certain desires and emotions, made certain practical inferences, and performed an act of a certain type. But according to Fodor's account, she is led to make these attributions by what I shall call a cold methodology: one that chiefly engages our intellect and makes no essential use of our own capacities for emotion, motivation, and practical reasoning.

Suppose now that we rewrite the scene as follows: Hermia had been speaking intimately with Lysander when Demetrius walked past. She noticed Demetrius's giving a brief glance in their direction and then looking a strange way that she found very difficult to categorize or describe. That ineffable look, she explains, helped persuade her that Demetrius was likely to do harm to Lysander—that look, taken together with other evidence. I find it hard to see how a facial expression mimicked on one's own face, particularly an expression one finds difficult to categorize, might translate into a predicate one can plug into a Fodorian law.

In contrast, let us suppose Hermia were to employ a hot methodology throughout. She acquires information concerning Demetrius's situation, background, and behavior in broadly the same way as she picked up the hard-to-categorize look Demetrius cast in their direction: by mirroring, or simulation. She uses her own perceptual, cognitive, motivational, and emotive resources to mirror Demetrius's mind as best she can and then, continuing in the role of Demetrius, predicts what he will do by deciding what to do. This makes it easy to understand how the look she picked up in her own facial muscles, and the underlying emotion she picked up as a result, would work together with other evidence. The most important point is this: Hermia would not have to categorize her second-hand emotion at all. She would have only to use it, that is, to allow it to do its work, to influence her behavior—all safely off-line, of course, strictly in the role of Demetrius. That, in very broad terms, is the view I take (Gordon 1986, 1992), as do Alvin Goldman (1989, 1992), the developmental psychologist Paul Harris (1989, 1992), and a few other philosophers and psychologists.[3]

It is clear how we sometimes use this methodology to predict and explain our own behavior in hypothetical situations. To predict and explain the actual behavior of other agents, it often suffices to call on our own emotions, desires, and practical reasoning, with little or no modification: Demetrius will get in out of the rain just as we will, and for the same reasons. That is the default mode of simulation. But sometimes the default mode will not do the job, so we make adjustments, such as viewing the list of award winners through the eyes of one's competitor. How do we accomplish that?

To illustrate, let us turn from Hermia's understanding of Demetrius to that of the actor whose job it is to play Demetrius. Actors today typically aim to motivate from within what their character says and does. Whatever else that involves, it requires the actor to become emotionally and conatively engaged in the action. He is not allowed to do this any old way; he is supposed to have some understanding of the character's motivation. (It is common for a drama coach to ask student actors to explain their character's actions, and to do so in the first person: "Why did you do that?") Because the actor's very aim is emotional and conative involvement, it does not seem plausible that he would delegate the task of understanding Demetrius to a procedure that works entirely without such involvement on his part. It does not seem plausible that he would first try to derive a list of beliefs, desires, and emotions that best fits the evidence and then try to recreate these propositional attitudes in himself. It seems more plausible that the actor would employ a "hot" methodology from the start.

Here is the way it would work. Many actors speak of transforming themselves, of becoming the characters they play. This typically involves an imaginative shift in the reference of indexicals. There is a character in the dramatis personae who becomes in your imagination the referent of the pronoun "I," and his time and place become the referents of "now" and "here." One very perceptive actor, Ray McAnally of the Abbey Theater in Dublin, describes his thoughts while filming a scene in which he plays a future British prime minister: "I had a very interesting moment in 10 Downing Street, surrounded by pictures of all the previous Prime Ministers *and me in the middle of it*. And I realized it was true, I *was* the Prime Minister. It's not that I'm *pretending* to be the Prime Minister ... I *am* the Prime Minister" (*New York Times*, January 15, 1989). Of course, he is in fact pretending to be the PM, only he is doing it so well that he is oblivious to that fact. (By the way, he is not pretending that the following counteridentical is true: "Ray McAnally is the PM." Analogously, although I may pretend on July 1 that "it is now New Year's Eve," I am not pretending that *July 1* is New Year's Eve.)

Simulation permits us to extend to others the modes of attribution, explanation, and prediction that otherwise would be applicable only in our own case. One reason I emphasize transformation is that a number of writers, both pro and con the simulation theory, take simulation to involve an implicit inference from oneself to others. It is essentially the old argument from analogy, which requires that you first recognize your own mental states, perhaps under certain imagined hypothetical conditions, and then infer that the other is in similar states. Although an actor will sometimes ask, "What would *I* do, or feel?" that is quite different from transforming himself into the character he is playing. McAnally tells us: "I can play any part in the world mentally, but I must always struggle to get Ray McAnally out of the way, with his opinions, his views, his background. They're my enemy."

The imaginative shift in the reference of indexicals reflects a much deeper, more important shift. Many of our tendencies to action or emotion appear to be specially keyed to an egocentric map. What triggers the action or emotion is the lion coming toward *me*, the meeting I am supposed to be at *now*, the insult directed to *me*, the award given to *my* child. (I suppose the map to include temporal as well as spatial relations, and even what we might call personal relations: for example, the insult was directed to *me*.)

What the actor can do is to recenter his egocentric map. Think of one of those transparent overlays on a map, with concentric circles showing the distance from any point you center it on. The actor can shift his egocentric overlay until it is centered on a particular character, place, and time, rather than on, say, Ray McAnally and his place and time. Tendencies to action or emotion that are egocentrically keyed now get triggered by a different set of objects and events. The lion approaching Demetrius is perceived as an *approaching* lion, a lion approaching *me*. Hence it is likely to evoke a flight response. An insult to Demetrius is an insult to *me*, and therefore likely to feel insulting and provoke a corresponding response. And so forth. In this way the actor becomes emotionally and motivationally engaged in the action.

To imagine being in the character's situation is often all the actor needs to motivate the appropriate action from within, but sometimes it is not enough. What the character says and does is radically different from anything the actor might in real life be inclined to say and do under similar circumstances. Suppose that the script stipulated that Demetrius actually does kill his rival Lysander. Even if the actor playing Demetrius did not have to perform the action (the killing takes place offstage), to become Demetrius he would probably try to see killing a rival as a genuine, psychologically possible option. This might require a bit of stretching, or moving away from one's own real-life character. The facts may have to be adjusted. One might pretend, for example, there is evidence that Lysander intends to kill *me*, or norms and values might be adjusted so that honor is seen as demanding that one kill a rival. To fine-tune the motivation may require a series of thought experiments, using what I have called hypothetico-practical reasoning (Gordon 1986).

Some Reasons to Think We Are Simulating

I noted that if Hermia herself were to employ a hot methodology, predicting what Demetrius will do by deciding (in the role of Demetrius) what to do, she could readily exploit emotion picked up second-hand by facial mimicry. To use them to help her predict Demetrius's actions, she would not have to categorize the emotion or call on any knowledge of lawlike relations between emotions, desires, beliefs, and actions. The same applies to empathetic responses to vocalizations and other so-called expressions of emotion. This capacity to integrate empathy of various sorts is an obvious advantage of a hot methodology and one reason to think it is what nonactors standardly use to explain and predict the actions of others.

Another reason is economy of storage. Again consider actors. How is the actor's knowledge of the roles she plays—consider especially the repetory actor who has to switch roles from one day to another—likely to be represented in her brain? It would be most extravagant if the actor had to store in memory an inventory of each character's long- and short-term mental states: all of the character's beliefs, all of the character's desires, and so on. A far more economical arrangement would be to store only a set of operations: just the set of changes or adjustments required for a given actor to "become," to transform herself mentally into, a given character. (Information about each character would thus be stored as a procedure rather than in declarative form, and specifically as a transformational procedure. This is comparable to a common method of compressing video signals, in which only the difference between each frame and its predecessor is represented in digital code rather than each frame in its entirety. Thus, a frame is treated as a set of changes from the preceding frame. One of the important disanalogies is that in the case of video signals, at least the initial frame must be coded in its entirety, whereas the actor would have no need to store information about herself.) Generalizing, the various people we are acquainted with would be represented in their mental aspects in the same way, as sets of transformational operations. Unlike the actor, of course, we are not called upon to transform ourselves audibly or visibly; we keep it to ourselves.[4] And whereas the actor transforms herself into characters in fictional worlds, we typically match our transformational operations with particular bodies in the actual world. Nevertheless, considerations of economy, among others, suggest that both the actor's knowledge of fictional roles and our knowledge of actual minds belong in the category not of speculative knowledge but rather of know-how: more specifically, of knowing how to transform ourselves into others.

I do not deny that we use generalizations from time to time, particularly generalizations about the long-term preferences and dispositions of particular individuals or cultures. But these function, I have suggested (in

Gordon 1992 and elsewhere), essentially as heuristics or rules of thumb. We must still keep our own decision-making system in readiness to fill in the implicit ceteris paribus clause—to warn us, for example, when a generalization yields a crazy prediction. Thus, we are better predictors than our generalizations are.

According to the simulation theory, until children know how and when to make imaginative adjustments in the factual basis for decision making when simulating others, they will be subject to certain errors in explaining and predicting the behavior of others. Their ability to explain or predict will be impaired by their inflexibility regarding the factual basis on which other people decide what to do. Among the limitations is that they will always predict as if any of the facts available to them were also available to the other. There is now considerable evidence that young children are subject to this impairment. Consider the following puzzle: Maxi puts his candy in location A, but while he is out of the room (playing outdoors) someone else relocates it to B. Where will Maxi go to get the candy when he gets back: to A or to B? You will probably answer, "To A, the place where he left it," for it is reasonable to assume that Maxi would not know that the candy had been relocated. But it would not have been your answer when you were a three year old. Until children reach about the age of four, they nearly always indicate the wrong place—the place where the candy is currently located. The results of a large number of experiments indicate that only at about age four do children correctly predict the behavior of people who, like Maxi, are not in a position to know some of the relevant facts. (If we accept the theory view we might try to account for the results by stipulating that the younger children have acquired, or are able to put to use, only a part of folk psychological theory, a part that does not include beliefs. But this seems ad hoc. The simulation theory, on the other hand, requires no special stipulation to account for the experimental results.)

A related consequence is that if pretending is the key to making and understanding mental state ascriptions, as the simulation theory holds, then a developmental pathology that severely restricts the capacity to pretend should also severely restrict a child's capacity to make and understand such ascriptions (even if in other respects her intelligence is close to normal). Autism is in fact such a pathology. The impoverishment of pretend play, particularly the absence of role play, is well known. In addition, people with autism are often said to treat people and objects alike: they fail to treat others as subjects, as having points of view distinct from their own. When a version of the candy relocation puzzle was administered to autistic children of ages six to sixteen, almost all gave the wrong answer—the three year old's answer. The evidence is that people afflicted with autism generally remain in the rigidly factive stage in explaining and predicting the behavior of others. (Although many of them are also mentally retarded, those tested were mostly in the average or borderline IQ range. Yet children with Down's syndrome, with IQ levels substantially below that range, suffered no deficit; almost all gave the right answer. See Gordon and Barker 1995 for a discussion of the philosophical significance of some of the data concerning autism.)

I am not denying that one could construct ad hoc versions of the "theory" theory, or, more broadly, a cold methodology theory, that would have one or another of these consequences. My point is that the simulation theory, to its credit, is stuck with these consequences.

Reversion to Emotional Contagion and Online Decision Making

If my account of the uses to which we put facial empathy is correct, then it is by a kind of "contained" emotional contagion that we recognize the emotional expressions of others. The emotion one "catches" from another gets referred to the individual from whom one caught it and (even if one has no label for it, as I imagined in the case of Hermia and Demetrius) integrated with other information about the same individual. But it is not enough just to refer the emotion to its source; there must be a containing mechanism that segregates these emotions from those arising out of our own perceptions and memories. The mechanism must allow the former to affect the decisions we make in simulating the other but prevent them from affecting our own decisions and actions in the way that our own "original" emotions would.

If this is true, then, as I said at the beginning of this chapter, the mirroring of minds is more pervasive than even Hume had thought. And perhaps most important, it takes a special containing mechanism to keep the emotion-recognition process from reverting to ordinary emotional contagion. This is important because unless we assume, implausibly, that the mechanism is absolutely fail-safe, it follows that the process does sometimes revert to ordinary contagion. And that is significant because ordinary emotional contagion, of the sort to which even six-month-old infants are subject, transmits genuine, even if second-hand, emotion. When another person feels joy, embarrassment, grief, or the emotional aspects of physical pain, contagion gives us the same emotion, sometimes (as in social referencing) toward the same "object." The second-hand emotion has the usual physiological manifestations, such as those mediated by the autonomic nervous system, just as if it had originated in us, and it affects our decisions and actions in the way that "first-hand" emotions do. To mention only one biologically important example, a parent or caregiver, and sometimes even an unrelated bystander, will tend to be moved to relieve the infant's pain, in the way that one's own pain moves one to find relief. At least in the case of negative emotions, second-hand emotion produced by ordinary emotional contagion is something like an enfranchised political representative; it becomes a voting member of one's own motivational system, and thus interpersonal conflicts will tend to become intrapersonal (to borrow from Richard Hare).

If the simulation theory is right, however, this is only part of the story. The decision making we perform in simulating another must itself be segregated from our own decision making; it must be decoupled from the mechanisms that ordinarily translate decision making and intention formation into action. Here, too, only a thin line will separate one's own mental life from one's representation of the mental life of another. Just as contained emotional contagion may become true contagion, our off-line decision making may go online and have the behavioral consequences of ordinary decision making. Children, especially, but also grownups, sometimes have a hard time segregating pretense from reality. But if pretense and reality get blurred in the case of counter-indexical pretending, or role taking, then we should expect a quite distinctive and weird sort of error: a blurring of the distinction between oneself and the other. In fact, this is not uncommon among actors, who are often known to carry their roles into their private lives. Special vigilance is required to keep oneself from being corrupted by an evil role—as the young actor who played Amon Goeth, the Nazi commandant in the movie *Schindler's List*, learned. In an interview in the *New York Times* ("Self-Made Monster: An Actor's Creation," February 13, 1994), Ralph Fiennes admitted that after living with Goeth for so long, he found himself sympathizing with him. "In the end [the character] becomes an extension of your own self. You like him. . . . You feel sort of peculiar because you might have at moments enjoyed it and at the same time you feel slightly soiled by it." In one scene, Goeth fires his rifle from his balcony overlooking the camp at prisoners picked out at whim. Fiennes said that he prepared for the scene by recalling "that boyish thrill with an air rifle when you aim at cans on a wall. That satisfaction when you hit a target—it gives you a kick." It is as basic, he says, as standing in front of a windowpane as a child and seeing how many flies he could smash with his hand. With this, the reporter notes, "He started to smash the table with an open palm but checked himself and brought it down gently."

Members of the audience, too, and indeed readers of novels, often take the roles of various characters in their attempt to understand what is going on, and audiences respond empathetically to the actors' facial expressions. The simulation theory has the consequence that people ought to be prone to such confusion in nontheatrical settings as well, for there too our transformational representations of other people will have a tendency to go online.

Ethical Consequences

In addition to its epistemological role in predicting, understanding, and explaining conduct, putting oneself in the other's place appears to play a prominent role in ethical evaluation. It is a procedure we consciously use in everyday moral thinking. It appears essential to the application of reciprocity principles such as the Golden Rule, and it is arguably needed also for the application of a principle of universalizability. That is a heavy load for one mental procedure to bear. As may be expected, a procedure suitable to one of these tasks may not be suitable to another. I have emphasized that simulation of the sort we employ in explaining or predicting another's behavior may involve transforming oneself psychologically into the other person. But is that what we want in moral assessment? The actor McAnally avows that he always bears in mind Aquinas's dictum that nothing is ever chosen except under the aspect of good. "My job in life," he says, "is to see the good aspect of what the character is choosing." This, he says, makes errant behavior feel perfectly logical. Surely that is not what we want. To see the other's actions as performed "under the aspect of good," as subjectively justifiable or at least excusable, may sometimes be a useful step in forming a moral assessment, but it is obviously not where we want to end up. Unlike the explanation or prediction of behavior, moral assessment requires holding back.

Compare the task of giving someone practical advice—prudential advice or just technical advice. We naturally preface advice with the phrase, "If I were you . . ." This is more than a figure of speech. Typically when we set ourselves up to give advice, we imaginatively project ourselves into the person's problem situation.[5] The reason is fairly obvious: we try to conceive the task as a practical one, one that calls for our action, so that we can marshal our own practical know-how or expertise, or at least our independent practical judgment. That is what we usually call on when giving practical advice.

Although we imaginatively project ourselves into the person's problem situation, it is always important in giving advice to hold back in certain ways from identification with the other person—that is, from making the further adjustments required to imagine being not just in that person's situation but that person in that person's situation. Otherwise, we lose the very advantages that make our advice worthwhile: the special know-how or the independent judgment. In certain cases the problem situation itself includes some of the other's psychological states, and in those cases what we offer is conditional advice: "Given your belief [preference, etc.], what I would do is . . ." Nevertheless, we must hold back in other ways if our advice is to be of any use.

The same point—that it is always important to hold back in certain ways from identification with the other person—appears to hold true in moral assessment. Indeed, a similar point is implied in Adam Smith's chief criticism of Hume's account of sympathy. According to Smith, rather than simply respond to another's pleasure with pleasure and to another's suffering with suffering, we find ourselves turning our attention to the cause of the other's emotion. We imagine ourselves being in the other's situation, ourselves faced with whatever is causing the other's emotion. Then, in imagination, we respond independently, in our own way, to the imagined cause: "We either approve or disapprove of the conduct of another man according as we feel that, when we bring his case home to ourselves, we either can or cannot entirely sympathize with the sentiments and motives which directed it" (Smith 1790/1976). If we find the same sentiments and motives evoked in ourselves, we approve of the conduct; if not, we disapprove.

Unfortunately, Smith makes it appear that we use the identical method to psych out what the other's sentiments and motives are: "As we have no immediate experience of what other men feel, we can form no idea of the manner in which they are affected, but by conceiving what we ourselves should feel in the like situation" (Smith 1790/1976). Here Smith misses the distinction I made earlier between two kinds of simulation: between just imagining being in X's situation, and making the further adjustments required to imagine being X in X's situation. Adding that distinction, we get the following decision procedure: imagine being in X's situation, once with the further adjustments required to imagine being X in X's situation, and once without these adjustments. If your response is the same in each case, approve X's conduct; if not, disapprove.

There is at least one problem, though: we sometimes disapprove of our own conduct—indeed, even our present conduct. So we should not use our own conduct, as we respond in imagination to X's situation, as the standard for judging X's conduct. We should use as our standard only hypothetical conduct of our own that we approve of. But how, it might be asked, could Smith's procedure yield disapproval of one's own conduct? Smith expressly addresses this question. Just as with another's conduct, he asserts, we disapprove of some of our own conduct if we cannot entirely sympathize with the sentiments and motives that directed it. But how can we fail to sympathize with our own present sentiments and motives? Again Smith has an answer:

> When I endeavour to examine my own conduct ... I divide myself, as it were, into two persons; ... I, the examiner and judge, represent a different character from that other I, the person whose conduct is examined into and judged of. The first is the spectator, whose senti-

ments with regard to my own conduct I endeavour to enter into, by placing myself in his situation, and by considering how it would appear to me, when seen from that particular point of view. The second is the agent, the person whom I properly call myself. (Smith 1790/1976)

Obviously, when simulating a person other than myself, I can in that role be out of sympathy with myself. But how can this happen when I myself, with my self-same emotions and motivation, am the spectator? It can happen only if the very process of becoming a spectator, and of coming to regard myself as an other, changes me. I believe that the process does change us, and that this was Smith's contention, too. But the topic is a complex one: observing myself trying to argue the case in a paragraph or two, I see a lot of hand-waving. So I will not try.

Notes

1. I base this account on Meltzoff and Gopnik (1993). The quotation from Darwin appears in that chapter.
2. Perhaps we should amend Darwin's description of his son's sympathetic emotion. The infant's response to the nurse's pretend grief may not have been grief but rather just sadness, for grief, if I am not mistaken, is always grief about something, whereas sadness, a mood, need not be about anything in particular. See Gordon (1987).
3. For essays for and against the simulation theory by these and other authors, see Davies and Stone (1995a, 1995b).
4. Actors, of course, do not keep their representations of other mental lives entirely off-line, for they have to act. Many contemporary actors emote inwardly, not just to understand what their character is doing and why but also in order to emote outwardly, for the audience. (But, as suggested earlier, an inward emotion may also be induced by the outward display of emotion).
5. But we are not committing ourselves as to what in fact we ourselves will do when such a situation arises. In, "If I were you, I would X" the second occurrence of "I" does not refer to the speaker. Consider, for example, "If I were you, I would stay away from *me*."

References

Darwin, Charles. 1896. *The Expression of the Emotions in Man and Animals*. New York: Appleton Press.

Davies, Martin, and Tony Stone, eds. 1995a. *Folk Psychology: The Theory of Mind Debate*. Oxford: Blackwell.

Davies, Martin, and Tony Stone, eds. 1995b. *Mental Simulation: Evaluations and Applications*. Oxford: Blackwell.

Fodor, Jerry A. 1987. *Psychosemantics: The Problem of Meaning in the Philosophy of Mind*. Cambridge, Mass.: MIT Press.

Goldman, Alvin I. 1989. Interpretation psychologized. *Mind and Language* 4:161–185.

Goldman, Alvin I. 1992. In defense of the simulation theory. *Mind and Language* 7:104–119.

Gordon, Robert M. 1986. Folk psychology as simulation. *Mind and Language* 1:158–171.

Gordon, Robert M. 1987. *The Structure of Emotions: Investigations in Cognitive Philosophy.* Cambridge: Cambridge University Press.

Gordon, Robert M. 1992. The simulation theory: Objections and misconceptions. *Mind and Language* 7:11–34.

Gordon, Robert M., and John A. Barker 1995. Autism and the "theory of mind" debate. In G. Graham and L. Stephens, eds., *Philosophical Psychopathology: A Book of Readings.* Cambridge, Mass.: MIT Press.

Harris, Paul. 1989. *Children and Emotion.* Oxford: Blackwell.

Harris, Paul. 1992. From simulation to folk psychology: The case for development. *Mind and Language* 7:120–144.

Hume, David. 1739/1978. *A Treatise of Human Nature.* 2d ed. Edited by L. A. Selby-Bigge, with text rev. and variant readings by P. H. Nidditch. Oxford University Press.

Meltzoff, Andrew, and Alison Gopnik. 1993. The role of imitation in understanding persons and developing a theory of mind. In S. Baron-Cohen, H. Tager-Flusberg, and D. J. Cohen, eds., *Understanding Other Minds.* Oxford: Oxford University Press.

New York Times. 1989. A living definition of acting. January 15.

New York Times. 1994. February 13.

Smith, Adam. 1790/1976. *The Theory of Moral Sentiments.* Edited by D. D. Raphael and A. L. Macfie. Oxford: Clarendon Press.

Chapter 10
Simulation and Interpersonal Utility
Alvin I. Goldman

Interpersonal Utility and Moral Theory

The aim of this chapter is to show how research in cognitive science is relevant to a certain theoretical issue in moral theory: the legitimacy of interpersonal utility (IU) comparisons. The legitimacy of IU comparisons is important to ethics for at least two reasons. First, certain ethical theories rely essentially on IU judgments, so the viability of those views rests on the tenability of such judgments. Repudiation of IU comparisons threatens not only utilitarianism but all other approaches to social theory that invoke welfare comparisons across individuals. Second, skepticism about IU comparisons has led social choice theorists to adopt conceptual tools unsullied by such comparisons, but the normative adequacy of these tools is in doubt. I specifically have in mind the concept of Pareto optimality, or Pareto efficiency. An allocation of goods is called Pareto efficient just in case there is no feasible alternative allocation where everyone would be at least as well off and at least one agent would be strictly better off. This definition invokes only intrapersonal welfare comparisons. But the concept of Pareto efficiency is extremely weak as a normative concept. Pareto-efficient states are a dime a dozen, and there is little temptation to regard every such state as normatively acceptable. (For example, an allocation whereby one agent gets everything in the economy and all others get nothing is Pareto efficient, since any reallocation would make the first agent worse off.) It is therefore essential to review the question of whether principled avoidance of IU comparisons is indeed required.

As Amartya Sen points out, different moral theories may require different types of IU comparisons.[1] Classical utilitarianism, for example, requires the maximization of a welfare sum, and therefore needs comparisons of welfare differences across individuals. On a Rawlsian theory, by contrast (at least the economists' version of Rawls), what is to be maximized is the

This chapter originally appeared in *Ethics* 105 (July 1995): 709–726. It is reprinted with the permission of the University of Chicago Press.

welfare of the worst-off individual. This requires IU comparisons of only welfare levels, not welfare differences. In my discussion, however, this variety of types of IU comparisons will play a minor role. I shall mainly address the possibility of, and the basis for, IU comparisons in a more general and unspecific form.

Some ways of trying to legitimate IU judgments strike me as unsatisfactory. First, writers such as Lionel Robbins attack IU comparisons as being devoid of descriptive meaning but allow them as (disguised) normative statements or expressions of value.[2] I assume, however, that the main question of interest is the descriptive one: Are there types of (relational) facts that IU comparisons purport to capture? Moral and social theorists typically appeal to IU relations as a factual ground for normative judgments. If IU judgments are themselves purely normative, this factual ground disappears.

Similar reservations apply to purely subjective and dispositional interpretations of IU comparisons. For example, one might interpret IU judgments as purely personal statements, describing how someone feels when answering a question of the kind, "Do I feel I would be better off as person I in social state X rather than person J in social state Y?" or "Which position would I choose if offered the opportunity to exchange places with one of them?"[3] If the judge says (or thinks), "I feel I would be better off as person I in social state X rather than as person J in social state Y," then as long as the statement accurately reflects how he feels, his IU judgment is correct, for the judgment only purports to describe his feelings (or the choice he would make if given the opportunity). The obvious problem with this subjective interpretation is that apparently conflicting IU judgments by different judges do not genuinely conflict on this interpretation, for each concerns only the speaker's feelings or choices. This hardly captures what is normally intended by IU judgments.

A more subtle approach would be a dispositional, or "secondary quality," approach. This would say that IU relations consist in nothing more than propensities on the part of two people's situations to evoke IU feelings (of the kind described above) in judges who contemplate the matter. Problems arise, however, when different judges are disposed to react differently to the same two people's welfare positions. Should some judges be deemed more authoritative than others, and on what grounds should they be so deemed? Although the dispositional approach has some prospects for further development, I shall not pursue it here. Most theorists who wish to use IU comparisons in ethics intend them in an objectivist and nondispositional (that is, nonreceptivist) fashion. If there are no "God's-eye" facts about relative welfare, the attractiveness of IU comparisons for moral purposes is greatly reduced. Hence, I shall construe the question about the legitimacy of IU comparisons as one of whether such comparisons are objectively viable.[4]

A further preliminary problem concerns the notion of utility, or welfare. What kinds of states of individuals are we comparing when we make interpersonal utility comparisons? Broadly speaking, there are two approaches to the concept of utility: the hedonic and the preference-satisfaction approaches. The hedonic approach thinks of utility as involving enjoyment, pleasure, contentment, or happiness. The preference-satisfaction approach thinks of utility as involving the realization or actualization of preferences or desires. Nineteenth-century utilitarians such as John Stuart Mill and Sidgwick favored the hedonic approach, while economists in this century have opted for preference satisfaction. This switch occurred largely because preference has a positivistic respectability: there appears to be an operational test for determining what a person prefers, that is, give her a choice and see what she does. On the other hand, it just is not clear that satisfaction of (actual) preferences captures our commonsense conception of well-being, or what is good for a person. In the case of social policies, for example, we do not suppose that whatever policy a person prefers is better for him. He might simply be mistaken about the consequences of competing policies and thereby prefer the policy that is actually worse for him.[5] There are other complications as well in trying to choose between the two approaches and make pertinent refinements.[6] These details cannot be pursued here, however, for they would take us far afield. Fortunately, I do not think we are forced to choose between the approaches or examine the requisite refinements, because the main problems and prospects for IU comparisons (at least the ones I shall discuss) arise for either approach and do not depend on refinements. Thus, I shall not firmly choose between the two types of utility conceptions, though most of my discussion will highlight the hedonic conception.

Interpersonal Utility, Philosophy of Mind, and Cognitive Science

When people inquire about the legitimacy of IU comparisons, there are two main questions, or families of questions, they have in mind: semantic and epistemic questions. The semantic questions include the following: Are IU judgments genuinely meaningful? Do statements such as "Person X enjoyed last night's party more than person Y enjoyed it" have any factual sense or content? Or (in a more metaphysical vein), are there any "facts of the matter" that make such statements true or false? The epistemic questions include the following: Do we (or could we) have epistemic access to (alleged) facts of this sort? Can we know or justifiably believe that any such statements are true? Both sorts of questions, I believe, must be answered in the affirmative if IU judgments are to be granted legitimacy. Mere meaningfulness, for example, would not suffice for the purposes of moral and social theory. Unless IU comparisons can be known, or justifiably believed, there is little basis for using them in policy-making. But it is also important to distinguish between the two sorts of questions. Much of the discussion of IU judgments (especially by economists) has been conducted within a positivistic or quasi-positivistic framework, according to which knowability or verifiability is essential for meaningfulness. This framework should be rejected; we should not conflate meaningfulness with knowability. Nonetheless, both are of interest.

Let us begin with meaningfulness. To assess the possible meaningfulness of IU judgments, we must turn to the philosophy of mind. IU comparisons are comparisons of mental states (either states of contentment or states of preference), so if such comparisons are meaningful, their meaning must derive from the way that meaning gets attached to mental language in general.[7] What does philosophy of mind have to say on this subject?

I shall consider two approaches: experientialism and functionalism. Experientialism is the traditional view that mental language gets its meaning, primarily and in the first instance, from episodes of conscious experience of which an agent is more or less directly aware. Experientialism says that felt (so-called occurrent) desires and felt episodes of enjoyment or contentment are genuine psychic magnitudes; they are experientially or phenomenologically real, and the meanings of the relevant mental predicates that describe them arise from points or intervals on the experiential scales that these predicates denote. Although each person has direct introspective access only to his or her own phenomenological scale, the meaning of "delighted," for example, is generalized to a point or range on anybody's enjoyment scale. It remains an open question whether experientialism can accommodate the possibility that one person, A, could know that another person, B, has an experience at a point on the latter's enjoyment scale that corresponds to A's own "delighted" range. But this epistemological question is a separate one. As far as meaning is concerned, experientialism seems to accommodate IU comparisons quite comfortably.[8]

I turn next to functionalism, which has been the most popular theory in the philosophy of mind during the last twenty or thirty years. Insofar as we are concerned with semantic questions, the relevant version of functionalism is analytic (or commonsense) functionalism. This doctrine says that the meaning of each mental predicate is specified by a distinctive niche in a network of commonsensically known causal laws, laws relating stimulus inputs, internal states, and behavioral outputs. For example, "thirst" means (roughly) "a state that tends to be caused by deprivation of fluid and to cause a desire to drink." All mental predicates get their meaning from the input–internal state–output laws in which they figure, and from no other source. Functionalism, I shall argue, has two problems. First, it is doubtful whether it can adequately capture IU judgments, especially the full range of IU judgments. Second, on entirely independent grounds, it is a problematic, unsustainable theory.

Before exploring functionalism's prospects for giving meaning to IU comparisons, let us review some of the reasons for people's skepticism about such comparisons. Here is a typical statement of skepticism by the economist Robbins (coupled with the point that IU comparisons are not needed by economic science):

> It is one thing to assume that scales can be drawn up showing the *order* in which an individual will prefer a series of alternatives, and to compare the arrangement of one such individual scale with another. It is quite a different thing to assume that behind such arrangements lie magnitudes which themselves can be compared. This is not an assumption which need anywhere be made in modern economic analysis, and it is an assumption which is of an entirely different kind from the assumption of individual scales of relative valuation. The theory of exchange assumes that *I* can compare the importance *to me* of bread at 6d. per loaf and 6d. spent on other alternatives presented by the opportunities of the market. And it assumes that the order of my preferences thus exhibited can be compared with the order of preferences of the baker. But it does *not* assume that, at any point, it is necessary to compare the satisfaction which *I* get from the spending of 6d. on bread with the satisfaction which *the Baker* gets by receiving it. That comparison is a comparison of an entirely different nature. It is a comparison which is never needed in the theory of equilibrium and which is never implied by the assumptions of that theory. It is a comparison which necessarily falls outside the scope of any positive science. To state that A's preference stands above B's in order of importance is entirely different from stating that A prefers *n* to *m* and B prefers *n* and *m* in a different order.... It has no place in pure science.[9]

Another classical basis for skepticism about IU comparisons is the von Neumann–Morgenstern treatment of utility theory.[10] Within this theory, numerical assignments to a person's preferences or desires are unique only up to a linear transformation. In other words, numerical assignments are purely arbitrary as to the origin (zero point) and unit selected. Numerical assignments to any other person's preferences or desires are equally arbitrary with respect to origin and unit. Hence, there is no principled or meaningful way in which the strength of one person's preferences can be put into correspondence with the strength of a second person's preferences.

Does functionalism provide any resources for transcending von Neumann–Morgenstern utility? At first blush, no. After all, functionalism is just a fancified version of behaviorism, and it appears that its only resources for assigning utilities are actual choice behavior and tendencies to cause possible choice behavior. But these are precisely the sorts of resources that utility theorists standardly invoke, and they see no prospect for extracting anything stronger than von Neumann–Morgenstern utility from these factors. However, the prospects for IU comparisons under functionalism may be a little better, especially if we consider the hedonic rather than the preference-satisfaction approach to utility. We should observe, first, that functionalism is not restricted to intentional choice behavior in measuring utility. There is also nondeliberate behavior like smiles, frowns, and other expressions of pleasure and displeasure to which functionalism might partly appeal in measuring utility (especially construed hedonically). If Smith is (spontaneously) smiling while Jones is (spontaneously) frowning, doesn't functionalism permit us to infer that Smith's hedonic state involves more utility than Jones's? This is an instance of a special subclass of IU comparisons in which one agent's utility seems to be positive and another's negative. Intuitively, these seem to be the easiest sorts of cases for making IU comparisons, as Richard Brandt has observed, since they only require us to determine positions on opposite sides of an origin, without worrying about unit comparability.[11] Functionalism also permits appeals to input conditions, and these may help identify types of internal states that are transpersonally positive or negative. Mental states produced by electric shock, for example, presumably have negative hedonic value across all individuals.

Although these points are suggestive, they hardly provide secure or systematic resolution of doubts about IU meaningfulness. Consider smiles and frowns, for example. Is it so clear that a smile expresses positive pleasure? Might it not instead express a mere reduction or diminution in discomfort or displeasure? Thus, Smith's smiling would not necessarily signal a state of positive utility. Analogously, Jones's frown might express a loss or diminution of pleasure (as compared with a previous or expected state of his) rather than a quantity of displeasure, and so his state might still involve more utility than Smith's. Furthermore, even if functional relationships could securely mark off positive and negative hedonic ranges, permitting interpersonal comparisons where people are on opposite sides of the hedonic divide, this does not begin to address interpersonal comparability when two people's hedonic states are either both positive or both negative.

We need not settle functionalism's prospects for rationalizing IU comparisons because there are independent reasons to regard it as an unsatisfactory theory of the meanings of mental terms. First, it is not clear that there is a set of lawlike generalizations accepted by the ordinary folk (especially a single set accepted by them all), in terms of which mental concepts can be defined. When one tries to spell out in detail just what these (implicitly) believed laws might be, it becomes questionable that the ordinary folk really possess such laws.[12]

Second, it is hard to pin down a set of laws that could define the mental predicates for all and only the beings that have mentality. As Ned Block has expressed the matter, functionalism faces problems of excessive chauvinism and excessive liberalism.[13] Functionalist definitions are in danger of excluding minded creatures that ought to be included or of including systems that lack mentality and should be excluded. In the latter category is the problem of creatures that lack "qualia" entirely but are classed as having the same psychological states as human beings because they are functionally equivalent to us. A related problem is that of spectrum inversion. It seems possible for two people to have functionally equivalent but qualitatively inverted arrays of color experiences, so there are possible differences in mental state not capturable by functionalism. A parallel possibility threatens in the hedonic domain: two people might have functionally equivalent arrays of enjoyment states that differ in felt intensity. For the same level of behavioral expressiveness or demonstrativeness, one person's feelings might be felt more strongly than another's.

A third major problem for analytic functionalism is epistemological: it cannot satisfactorily account for self-knowledge or self-ascription of mental terms.[14] As I show elsewhere, a functionalist understanding of mental terms would require computational operations that are surely not normally executed in real time, and may even be unfeasible.[15] Thus, if we view the theory of folk mentalizing (the theory of how ordinary people understand and deploy mentalistic concepts) as a part of cognitive science (and this is how I do view the matter), then functionalism is just an implausible theory, on (roughly speaking) empirical grounds.

There are several difficulties. Consider the case of a morning headache. You wake up to a distinctive sensation state and immediately classify it as a headache. According to analytic functionalism, "headache" is understood in terms of states that cause it or are caused by it. To classify a state as a headache, one would have to determine what inputs preceded it or what outputs and/or other inner states followed it. But when you classify your morning sensation as a headache, you may not recall anything that might have caused it. Nor need you have yet performed any action, such as taking an aspirin, that might help you identify it as a headache. Is there some other internal state you identify that prompts the "headache" classification? Perhaps you notice a desire to get rid of the state. But surely you may identify the headache before you identify this desire. Furthermore, appeal to the desire simply transfers the difficulty to that state. How do you classify that token state as a desire to be rid of the initial state? More generally, according to functionalism, the type-identity of a token mental state depends on the type-identity of its relata, and the type-identity of many of these relata (other internal states) depends in turn on their relata. Complexity ramifies quickly, and there is a threat of combinational explosion: too many other internal states would need to be type-identified in order to identify the initial state.[16] Thus, the functionalist account of the meaning of mental state predicates cannot be right, for it cannot account for the indisputable fact that we do make first-person mental ascriptions very quickly and with substantial accuracy. Only an experiential account of the meaning of mental terms—an account that locates the meaning of these terms in intrinsic rather than relational properties of mental states—seems adequate to explain the facts of self-ascription.[17] Thus, experientialism is a superior (semantic) account of mentalistic terms than functionalism, and we have previously seen that it is viable as an approach to the meaningfulness of IU comparisons. Whether the epistemological problem of IU comparisons can also be solved under experientialism remains to be seen.

Simulation and the Epistemic Problem of IU Comparisons

The epistemological problem facing IU comparisons under experientialism is straightforward. If anyone tries to compare the hedonic states of persons A and B, there seems to be no sound epistemic route to both of their states. Person A can introspect her own experiential state but not B's, and vice versa. A third party can introspect neither of their states. Is there any alternate route to these states and their relative hedonic properties?

In his discussion of interpersonal utility, John Harsanyi offers the following account of how people (third parties) approach questions of interpersonal comparisons:

> Simple reflection will show that the basic intellectual operation in such interpersonal comparisons is imaginative sympathy. We imagine ourselves to be in the shoes of another person, and ask ourselves the question, "If I were now really in *his* position, and had *his* taste, *his* education, *his* social background, *his* cultural values, and *his* psychological make-up, then what would now be *my* preferences between various alternatives, and how much satisfaction or dissatisfaction would *I* derive from any given alternative?"[18]

This account is very similar to that offered by many commentators, according to Sen. Sen speaks of the thought experiment of placing oneself in the position of others and finds this thought experiment used in the ethical treatises of Kant, Sidgwick, Hare, Rawls, Suppes, and Pattanaik, among many others.[19] I agree in finding this a natural way to approach the task of making interpersonal comparisons, but I would go much further and regard this as the standard, or at least a very common, strategy for making third-person mental ascriptions in general. This is the so-called simulation or empathetic approach to third-person ascriptions that several philosophers and at least one psychologist have recently defended, including Robert Gordon, Paul Harris, and me.[20] This approach is partly inspired by historical treatments of sympathy and *Verstehen*, but it also aspires to psychological realism. That is, it is presented as a hypothesis within cognitive science, open to the kinds of theoretico-experimental demands appropriate to any other cognitive science hypothesis.

The basic idea behind the simulation theory is that the naive ascriber of mental predicates does not have to represent and deploy a sophisticated commonsense psychological theory (set of causal laws) to ascribe mental states to himself or others successfully. Every normal agent has a set of mental mechanisms for generating new mental states from initial ones— for example, a decision-making system that takes desires and beliefs as inputs and churns out decisions as outputs. Such mechanisms or systems might be employed in a derivative fashion to simulate and ascribe mental states to oneself and others. To use the simulation heuristic, one would feed pretend initial states into such a mechanism—the initial states of the targeted agent—and see what further states the mechanism produces. For example, to predict someone else's choices, one can feed simulated desires and beliefs into a decision-making system and allow it to generate the same choice it would produce given genuine desires and beliefs. If the ascriber, predictor, or interpreter can feign the initial states of the target agent accurately (on the basis of prior information or simulations), and if the mental mechanism operates equivalently (or sufficiently close to equivalently) on both genuine and feigned states, there is an opportunity for the output states generated by the system to mirror, or accurately reflect, the ensuing states of the target agent.

There is prima facie evidence that (certain) mental mechanisms do have the capacity to be run in this simulative fashion. For example, when we engage in hypothetical deliberation, the practical reasoning system seems to generate provisional choices from imagined goals and beliefs in the same way that it generates real choices from genuine goals and beliefs during ordinary, "online" deliberation. Similarly, imagined perceptual, cognitive, and emotional experiences that are fed into a pleasure- or displeasure-generating system might generate (surrogate) hedonic outputs in a fashion that substantially mimics its mode of generating full-blown hedonic states from genuine experiences. This is what seems to happen to a spectator, reader, or observer of a drama, movie, novel, or news account, who "imaginatively identifies" with the protagonist, hero, or victim.[21] Simulation can also facilitate backward inference from behavior to mental states. If you and a companion go to a movie together, you can use your companion's subsequent behavior—including such verbalizations as "I loved it" or "I hated it"—to infer the nature of her enjoyment state. You would try to identify an enjoyment state (or range of states) such that simulation of that state together with other presumed concurrent states of hers would produce the observed verbal and/or nonverbal behavior.

Given the possibility of simulation-based third-person mental ascription, it is a small step to IU judgments. In the movie case, for example, you could compare the state assigned to your companion by simulation with your own introspected state of enjoyment regarding the movie and arrive at an interpersonal comparison. In making hedonic comparisons between two people other than yourself, you would simulate each of them in turn and compare the resulting hedonic states. I do not mean to suggest that comparisons are always easy. Some pairs of hedonic states may not be readily comparable, such as displeasures associated with physical versus psychic pain. In addition to metaphysical indeterminacy, accuracy of simulation (especially comparative simulation) may be more difficult in some cases than others.[22] In general, however, the experiential approach, coupled with the simulation or empathetic methodology of third-person mentalistic ascription, renders IU judgments highly intelligible and unsurprising.[23] This approach should be construed, once again, as a tentative hypothesis, subject to further empirical investigation.

As a working hypothesis, I shall henceforth assume that the simulation heuristic is what we most commonly employ in making IU judgments. This leaves the epistemic question still before us: Can such judgments, arrived at by these means, constitute knowledge or justified belief? That obviously depends on what it is to know something, and what it is to have justified belief. Focusing on knowledge, let us consider a reliabilist theory of knowledge, and for illustrative purposes, let us concentrate on a deliberately simplified version of reliabilism: "S knows that p IF AND ONLY IF p is true, and S arrives at a belief in p by means of a reliable (highly truth conducive) cognitive process."[24] On this theory of knowledge, it is certainly possible that IU propositions should be known to be true (or false). For example, certain uses of the simulation heuristic might be reliable ways of arriving at IU beliefs. I say "certain uses" because it is obvious that ill-informed or sloppy simulation has no chance of being generally reliable. If you have little or no antecedent information about B's initial circumstances, the pretend states you feed into your attempted simulation of B will have little correspondence to *B*'s actual initial states, and you are unlikely to generate an output that corresponds to *B*'s actual enjoyment level. But well-informed and sensitive use of the simulation heuristic may indeed be reliable. So the possibility of knowing IU propositions, by means of simulation, seems to be genuine.

The Scientific Legitimacy of IU Judgments

Even if people are prepared to grant that a reliabilist theory of knowledge is correct (or on the right track), many will doubt that this can serve as an account of scientific knowledge. Indeed, the very fact that the reliabilist theory could license simulation- or empathy-based knowledge would be greeted by some as a demonstration that reliably caused true belief is too weak a condition for scientific knowledge. After all, empathy is a paradigm of a nonscientific, merely intuitive mode of belief formation (in a pejorative sense of "intuitive"). This is an important issue, because the central objection to IU comparisons by many skeptics and critics is their lack of scientific legitimacy. Why, exactly, is empathy or simulation a nonscientific mode of belief formation? What, in general, are the criteria for scientific knowledge or justified belief (as opposed to casual or garden-variety knowledge or justified belief)?

We have already agreed that simulation could be reliable, so mere reliability must not be a sufficient qualification for scientific legitimacy. Perhaps an additional requirement is publicity. Maybe scientifically legitimate procedures must be public, and simulation or empathy violates this constraint. But what exactly is meant by publicity? Does the publicity constraint require that scientific procedures occur in the public sphere, outside the heads of individual scientists? Simulation admittedly violates that test, but so do all the reasoning procedures that are presumably admissible components of scientific procedure.

A better interpretation of publicity might be replicability. Perhaps simulation appears unscientific because it fails this test, but simulation might be replicable. Well-informed, skilled deployers of the simulation heuristic might frequently, perhaps even regularly, arrive at identical or similar IU conclusions concerning the same target states. If they did, wouldn't the replicability test be passed by at least those specific IU judgments? Would they then qualify as scientific (and, if true, as specimens of scientific knowledge)?

I have considerable inclination to say that they would. Others, however, might hold out for a stronger constraint: a "higher-order" constraint. They might insist that simulation can yield scientific knowledge if it not only is reliable, but there is uncontroversially scientific evidence—evidence of a different sort—that supports this reliability. Is scientific support for simulational reliability forthcoming?

The prospects for simulational reliability depend on at least two factors. First, psychological systems must operate on feigned or pretend input states in the same way they operate on genuine states, at least to a close enough approximation. Partial evidence for simulational reliability, then, would consist of evidence that supports such a property. A recent paper by Gregorie Currie adduces such evidence for one specific subdomain of simulation.[25] Although this leaves open the question of whether comparable evidence can be found in other subdomains, it nicely illustrates the possibility of finding relevant empirical evidence.

Currie wishes to contend, as I do, that simulative techniques might achieve accuracy by exploiting the same mechanisms that operate on real (unfeigned) inputs to produce real outputs. He tries to show that this is indeed the case for the domain of mental imagery. He begins by arguing that mental imagery is the simulation of vision, that is, creating a mental image of X amounts to imagining—simulating or pretending—that one is seeing X. (Of course, there is nothing interpersonal in this variant of simulation, but it involves the same heuristic of one process's emulating or mimicking another.) He then adduces empirical evidence supporting the contention that visual imagery operates by using the visual system itself; more specifically, it uses the central but not the peripheral parts of the visual system:[26]

> Farah has shown that there are content specific interactions between imagery and perception, for example that imaging an H affects detection of visually presented Hs more than it affects detection of Ts. She concludes from this that imagery and perception "share some common representational locus of processing". There is also strong evidence that visual deficits due to cortical damage are paralleled by imagery deficits.... Thus there is an association between loss of color perception and loss of color in imagery; patients with manifest loss of color vision perform badly on tests that would seem to require imaging ("What color is a gherkin?" as opposed to "What color is envy?"). Patients with bilateral parieto-occipital disease are sometimes impaired in the localization of objects in space but have less difficulty identifying objects themselves; patients with bilateral temporo-occipital disease sometimes show the reverse pattern of impairment. Farah and colleagues examined a case of each kind and found that "the preserved and impaired aspects of vision ... were similarly preserved and impaired in imagery".... Perhaps the most striking evidence from cortical damage comes from the work of Bisiach and colleagues on unilateral neglect. Patients with right-parietal-lobe damage often fail to detect objects in the left visual field; Bisiach and Luzzatti showed that their descriptions of imagery

when asked to imagine being placed in a familiar location (the Piazza del Duomo in Milan) show systematic neglect of objects on the left side of the imagined scene.[27]

Of course, this is all addressed to the simulation of vision only. The generic case of simulation will be much harder to study, since far less is currently known about the neurophysiological mechanisms of (say) choice, preference, or enjoyment. The moral, nonetheless, is that the accuracy of simulation is in principle subject to empirical investigation.

The problem of interpersonal simulation, however, cannot be addressed by the foregoing kinds of investigation. Interpersonal simulation can succeed only if there are certain psychological homologies or similarities between simulators and simulatees. Could there be empirical evidence for such homologies? One possibility is an evolutionary approach, and this is currently being pursued by Roy Sorenson, who argues that natural selection explains the effectiveness of the method of empathy. Sorenson summarizes his idea as follows:

Stepping into the other guy's shoes works best when you resemble the other guy psychologically. After all, the procedure is to use yourself as a model: in goes hypothetical beliefs and desires, out comes hypothetical actions and revised beliefs and desires. If you are structurally analogous to the empathee, then accurate inputs generate accurate outputs—just as with any other simulation. The greater the degree of isomorphism, the more dependable and precise the results. This sensitivity to degrees of resemblance suggests that the method of empathy works best for average people. The advantage of being a small but representative sample of the population will create a bootstrap effect. For as average people prosper, there will be more average descendants and so the degree of resemblance in subsequent generations will snowball. Each increment in like-mindedness further enhances the reliability and validity of the method of empathy. With each circuit along the spiral, there is tighter and tighter bunching and hence further empowerment of empathy. The method is self-strengthening and eventually molds a population of hypersimilar individuals.[28]

The general hypothesis, then, is that natural selection has molded us into a population of hypersimilar individuals, which helps support the reliability of simulation. It would not be easy to get firm scientific evidence for this hypothesis, but it seems possible, and if accomplished it would strengthen the scientific case for the accuracy of simulation in IU comparisons.

I do not mean to imply that simulational reliability can be supported only by such highly theoretical considerations. On the contrary, some evidence can be gleaned much more directly, by seeing whether simula-tion-based predictions of behavior are accurate. If it can be settled that many predictions of intentional behavior do in fact use the simulation heuristic, these successful predictions will also have been correct in infer-ring the agents' (mental) decisions or choices. This already provides sup-port for psychological similarity. It leaves open the questions crucial for IU comparisons—intensities or strengths of desire or preference. But empiri-cally observed success at empathy-based predictions of behavior does go some distance toward supporting psychological isomorphism.

This discussion has proceeded under the premise that the scientific legitimacy of empathy-based IU judgments depends on empirical support for simulational reliability, including empirical support for the psycho-logical similarities that simulational reliability requires. John Harsanyi, however, claims that empathy-based IU comparisons are scientifically admissible even though they are and must be based on a nonempirical similarity assumption.[29] Harsanyi says that IU comparisons using empathy rest on what he calls the similarity postulate: the assumption that, once proper allowances have been made for the empirically given differences in taste, education, and other characteristics between me and another person, then it is reasonable for me to assume that our basic psychological reac-tions to any given alternative will be otherwise much the same. He insists that this is a nonempirical, a priori postulate but holds that such non-empirical postulates are common in science. He specifically compares it with a priori nonempirical criteria of theory choice, such as simplicity, parsimony, and preference for the "least arbitrary" hypothesis. If two individuals show exactly identical behavior, or show differences in observ-able behavior that have been properly allowed for, then it is an arbitrary and unwarranted assumption to postulate some further hidden and un-observable differences in their psychological feelings.

Not everyone will agree with Harsanyi that a nonempirically based postulate of psychological similarity is scientifically admissible. But it is certainly plausible to hold that ordinary people presuppose such a postu-late in making simulation-based judgments, including IU judgments. The problem, then, is this: while we await future empirical research that might scientifically confirm (or disconfirm) the sorts of properties and relations that could underpin simulational reliability, how shall we assess the epistemic status of simulation-based judgments? What is the epistemic respectability of such judgments in the interim, pending outcomes of the sorts of research adumbrated earlier? I close by suggesting that the proper analogy is with judgments or beliefs about grammar, which enjoy con-siderable epistemic respectability despite analogous dependence on non-empirical postulates (at least according to the dominant view of such judgments).

According to the nativist perspective on language learning, pioneered by Noam Chomsky, the child's language learning mechanism has a rich store of innate knowledge that constitutes a strong bias in favor of acquiring certain grammars and against acquiring others. The primary linguistic data to which the language acquisition device is exposed are equally compatible with an indefinitely large class of grammars, many of which depart significantly from the grammar that the child actually attains. Chomsky argues that if the child had a purely "empiricist" mind, that is, if it operated only under the sorts of constraints (such as simplicity) that an empirical scientist would use in choosing among hypotheses, then he or she would have no reliable way of selecting the right grammar.[30] So the child's language acquisition device must in fact have a more powerful or highly constrained set of learning principles built into it: assumptions or postulates that bias it toward certain grammars rather than others. But what is the epistemic status of (mature) children's or adults' intuitive grammatical judgments about their native language? Surely their judgments about whether a sentence is grammatical (at least for garden-variety sentences), or whether one sentence is a paraphrase of a second, are epistemically very respectable. Normally we would say that such judgments are clear instances of knowledge. Moreover, we remain inclined to say this even when we are convinced by the Chomskyan story ascribing to the language learner a set of nonempirical postulates about which grammars are to be expected or not to be expected. Thus, reliance on nonempirical postulates does not seem to be a fatal flaw or barrier to epistemic respectability.

Similarly, I would argue, the fact that the simulation heuristic presupposes a nonempirical postulate of psychological similarity or homology should not disqualify its products from epistemic respectability (even if they do not earn the status of "scientific" knowledge). Of course, we must also be persuaded that the simulation heuristic regularly gets things right; otherwise the parallel with grammatical judgments would certainly collapse. But mere reliance on an innate assumption of psychological homology should not in itself be a hindrance to epistemic respectability. Thus, simulation may be in a position to deliver the epistemic (as well as the semantic) goods, thereby clearing the way for moral theory and social policy to avail themselves of IU comparisons.

Notes

1. Amartya Sen, "Interpersonal Comparisons of Welfare," in *Economics and Human Welfare*, ed. Michael Boskin (New York: Academic Press, 1979).
2. Lionel Robbins, *An Essay on the Nature and Significance of Economic Science*, 2d ed. (London: Macmillan, 1935).
3. See Sen, "Interpersonal Comparisons of Welfare," pp. 186–187.

4. In rejecting a dispositional interpretation of IU judgments, I reject only accounts that invoke reactive dispositions of judges; I do not (at this juncture) reject accounts, such as functionalism, that try to spell out intensities of mental states in terms of the agents' dispositions.

5. For discussion of this point and further reflections on rival approaches to the concept of utility, see Allan Gibbard, "Interpersonal Comparisons: Preference, Good, and the Intrinsic Reward of a Life," in *Foundations of Social Choice Theory*, ed. Jon Elster and Aanund Hylland (Cambridge: Cambridge University Press, 1986).

6. For example, what if people prefer a certain outcome for altruistic reasons? Does realization of this outcome constitute a contribution to their own well-being or utility? For discussion arising from such issues, see ibid.

7. They are not just comparisons of mental states, at least in the case of the preference-satisfaction approach, because they also concern which preferences are actually realized. The theoretically problematic issues, however, devolve on the mental-state comparisons.

8. Wittgensteinians, of course, have long challenged the defensibility of experientialism, centering their critique on the private-language argument. I do not believe that this critique is well founded, but that is a very large issue that cannot be broached here.

9. Robbins, *Essay on the Nature and Significance of Economic Science*, pp. 138–139.

10. The standard text on this subject is R. Duncan Luce and Howard Raiffa, *Games and Decisions* (New York: John Wiley, 1957).

11. Richard Brandt, *A Theory of the Good and the Right* (Oxford: Oxford University Press, 1979), pp. 259–260.

12. See Stephen Schiffer, *Remnants of Meaning* (Cambridge, Mass.: MIT Press, 1987), pp. 28–40.

13. See Ned Block, "Troubles with Functionalism," in *Readings in Philosophy of Psychology*, ed. Ned Block (Cambridge, Mass.: Harvard University Press, 1980), vol. 1.

14. This epistemological problem for functionalism should not be confused with the epistemological problem for IU comparisons (even under experientialism, for example).

15. Alvin Goldman, "The Psychology of Folk Psychology," *Readings in Philosophy and Cognitive Science*, ed. Alvin Goldman (Cambridge, Mass.: MIT Press, 1993).

16. This is a selective summary of the arguments presented in ibid. In particular, it does not explore the difficulty of complying with the subjunctive character of the lawlike relations under functionalism. See the original article for a full presentation of the difficulties.

17. This is a further thesis that is defended in ibid.

18. John C. Harsanyi, "Morality and the Theory of Rational Behavior," in *Utilitarianism and Beyond*, ed. Amartya Sen and Bernard Williams (Cambridge: Cambridge University Press, 1982), p. 50.

19. See Sen, "Interpersonal Comparisons of Welfare," pp. 186–187, esp. n. 9. However, Sen utilizes this empathetic thought experiment as the basis for his subjective interpretation of IU comparisons that I discussed in the first section. This is not how Harsanyi intends it, I feel sure, nor is it so clear that it is the intended interpretation of the other cited ethicists.

20. See Robert M. Gordon, "Folk Psychology as Simulation," *Mind and Language* 1 (1986): 158–171, and "The Simulation Theory: Objections and Misconceptions," *Mind and Language* 7 (1992): 11–34; Paul L. Harris, "From Simulation to Folk Psychology: The Case for Development," *Mind and Language* 7 (1992): 120–144; and Alvin I. Goldman, "Interpretation Psychologized," *Mind and Language* 4 (1989): 161–185, and "In Defense of the Simulation Theory," *Mind and Language* 7 (1992): 104–119. These and other papers on the simulation theory will be reprinted in *Mental Simulation*, ed. Martin Davies and Tony Stone (Oxford: Blackwell Publishers, forthcoming).

21. What is the exact relationship between the vicariously felt experiences of the empathic spectator and the genuine experiences of a target agent (when empathic acts are successful)? I do not think that we can answer this in detail yet. It is analogous to the question of how visual imagery is related to genuine vision. In both cases the relationships can be fully identified only through research in cognitive science. Good progress is being made on the nature of imagery (as illustrated later in the chapter), and similar progress can be anticipated if comparable resources are devoted to the nature of vicarious feelings. For some discussion of research in this area, see my "Ethics and Cognitive Science," *Ethics* 103 (1993): 337–360, and "Empathy, Mind, and Morals," *Proceedings and Addresses of the American Philosophical Association* 66 (1992): 17–41. A particularly insightful illustration of how pleasure can be derived from imaginative identification with others is given by Adam Smith: "When we have read a book or poem so often that can no longer find any amusement in reading it by ourselves, we can still take pleasure in reading it to a companion. To him it has all the graces of novelty; we enter into the surprise and admiration which it naturally excites in him, but which it is no longer capable of exciting in us; we consider all the ideas which it presents rather in the light in which they appear to him, than in that in which they appear to ourselves, and we are amused by sympathy with his amusement which thus enlivens our own." *The Theory of Moral Sentiments*, ed. D. D. Raphael and A. I. Macfie (Oxford: Oxford University Press, 1759/1976), p. 14. "Secondary" or derivative feelings are not restricted to interpersonally vicarious feelings. As Richard Moran points out, the person who says that it still makes her shudder just to think about her driving accident, or her first date, is exhibiting a response perfectly analogous to interpersonal empathy. See "The Expression of Feeling in Imagination," *Philosophical Review* 103 (1994): 78.

22. It would be a mistake to suppose that IU comparability is valuable for moral and social theory only if all pairs of states are both metaphysically and epistemically comparable. Even if only extreme differences are comparable, that might be helpful.

23. Not all simulation theorists accept experientialist accounts of the semantics of mental terms; in particular, Robert Gordon does not.

24. For expositions of reliabilism, see my "What Is Justified Belief?" in *Justification and Knowledge*, ed. George Pappas (Dordrecht: Reidel, 1979), reprinted in my *Liaisons: Philosophy Meets the Cognitive and Social Sciences* (Cambridge, MA: MIT Press, 1992), and *Epistemology and Cognition* (Cambridge, MA: Harvard University Press, 1986). For a recent permutation (on the theory of justified belief), see "Epistemic Folkways and Scientific Epistemology, " in *Liaisons*.

25. Gregorie Currie, "Mental Imagery as the Simulation of Vision," unpublished paper, Flinders University, Adelaide, Australia. 1994.

26. Currie draws heavily from an article by Martha J. Farah, "Is Visual Imagery Really Visual? Overlooked Evidence from Neuropsychology," *Psychological Review* 95 (1988): 307–313, which reviews various research, including her own.

27. Currie, "Mental Imagery," pp. 5–6.

28. Roy A. Sorenson, "Self-Strengthening Empathy: How Evolution Funnels Us into a Solution to the Other Minds Problem," unpublished paper, New York University, 1994, p. 1.

29. Harsanyi, "Morality and the Theory of Rational Behavior," pp. 50–51.

30. For a lucid exposition of these ideas, see William Ramsey and Stephen P. Stich, "Connectionism and Three Levels of Nativism," in *Philosophy and Connectionist Theory*, ed. William Ramsey, Stephen P. Stich, and David E. Rumelhart (Hillsdale, N.J.: Lawrence Erlbaum, 1991), especially sec. 2.2.

Chapter 11
Empathy and Universalizability
John Deigh

What makes psychopaths, as characterized in modern psychiatry and portrayed in literature and film, such fascinating figures? What is it about their evident lack of morality that transfixes our imagination as it chills our souls? To describe them, as they were once described, as moral imbeciles, does not even begin to convey the peculiarity of their condition. Though amoral, they appear nevertheless to be capable of reasoning, weighing evidence, estimating future consequences, understanding the norms of their society, anticipating the blame and condemnation that result from violation of those norms, and using these cognitive skills to make and carry out their plans. Some have been described as highly intelligent and socially adept, people whose gift for facile, ingratiating conversation can beguile even those already alerted to their pathology. The swiftness of their thought plainly does not fit with the ideas of stupidity and feeble-mindedness that the older description carries.

Nor are psychopaths well described as maniacs, pyro-, klepto-, homicidal, or otherwise. They are persistent wrongdoers, to be sure, but they are not or not necessarily driven to commit their misdeeds. No inner compulsion or violent emotion is essential to their disorder. Think of Richard Hickock as depicted in Capote's *In Cold Blood*, or Bruno C. Anthony brilliantly portrayed by Robert Walker in Hitchcock's *Strangers on a Train*.[1] Each is a smooth and bloodless operator, not someone subject to irresistible impulses, not someone governed by personal demons.

Think too of the contrast Capote sets up between Hickock and his partner in crime, Perry Smith. Smith, who wielded the knife and pulled the trigger in the murders for which he and Hickock were executed, is depicted by Capote as a man with a very weak grip on reality, given to dream and delusion, and also prone to explosive rage. In Capote's reconstruction of the crime, these factors combine to ignite Smith's murderous action, and the savagery of his conduct is made even more extraordinary by the prior

This chapter originally appeared in *Ethics* 105 (July 1995): 743–763. It is reprinted with the permission of the University of Chicago Press.

concern he shows for the comfort and, indeed, safety of his victims. Both during the commission of the crime and in the subsequent events leading to execution, Smith exhibits this soft side as well as a capacity for sympathy and connection. These are dispositions wholly absent in Hickock's character. Capote describes him as an exceptionally shallow man, always scheming, filled with petty emotions when his schemes fail, and incapable of deeper feelings either for himself or others. In particular, he feels no compunction about his actions, however wrong or injurious they may be, and exhibits no shame, regret, or remorse over them. The crime is his idea, and he befriends Smith because he sees in Smith's capacity for violence something very useful to forwarding his criminal ambitions. He, on the other hand, is not, despite his bluster, inclined to violence. And unlike Smith, he has no trouble keeping a clear head and a steady mind as he pursues his evil ends.

Of course, you might think there is nothing really peculiar about Hickock's character. He is like any number of common thieves, con artists, and gangsters who lead lives of crime, a vicious man who is interested only in himself and who cares little about others. To call him a psychopath is just to put him in that grab bag category psychiatrists use for any habitual wrongdoer or troublemaker who is not obviously psychotic. Yet this category, despite its overuse by many psychiatrists, can be more tightly constructed so that it applies to a distinct type of personality, a type defined by specific emotional deficits and imaginatively rendered in Capote's description of Hickock.[2] These deficits include the inability to enter into genuine friendships or to form attachments of loyalty and love, and also the lack of a capacity for moral feeling and emotion, the lack of a conscience. If one makes possession of these deficits a defining condition of the category, then the difference between a psychopath and your run-of-the-mill career criminal is clear. In the latter's life, unlike the former's, there are or can be people to whom the criminal is emotionally committed and in relations with whom he is liable to some experience of moral feeling, some stirrings of conscience. Thus even for as hard bitten and callous an outlaw as the one Humphrey Bogart played in *High Sierra*, a man known to police and prosecutors nationwide as "Mad Dog" Earle, there is an Ida Lupino to whom he is attached by bonds of trust and love and in relations with whom experiences of regret and remorse are possible. By contrast, in the life of a psychopath, on this tighter construction of the category, there can be no Ida Lupinos.

Tightening the category by making these emotional deficits essential to it, not only saves the term "psychopath" from becoming merely a fancy word for habitual wrongdoer, but also sharpens the issues of moral responsibility and criminal liability that it raises. For "psychopath" is not exclusively a term of clinical practice. It is also a term psychiatrists, criminologists, and other legal theorists use to pick out a class of criminals whose crimes result from severe mental disorder but who do not meet the law's standard tests for being insane. Those tests concern the agent's moral knowledge and his capacity for self-control, whether he knows the difference between right and wrong and whether he can conform his conduct to the limits the law imposes, and it is generally conceded that psychopaths, whatever the nature of their disorder, know right from wrong and have the requisite self-control. Consequently, whether they are morally responsible for their crimes is an issue on which opinion divides according as one believes wrongdoing that is the product of severe mental disorder is blameless, no matter how grave the offense, or believes wrongdoing that the agent knows or should have known is wrong and that he could have averted merits blame, no matter how screwy the agent. The psychopath, as a mentally disordered individual, thus puts pressure on our traditional criteria for moral responsibility, the more so, the more one can characterize the disorder as something other than a proclivity for wrongdoing.

Admittedly, the pressure it puts on the traditional criteria shrinks to nothing when those exerting it can produce no more compelling evidence of the disorder than the agent's persistent wrongdoing. In this case, the argument for excusing psychopaths from moral responsibility comes to little more than psychiatric ipse dixitism. The argument, however, gains considerable strength when one uses possession of certain emotional deficits to define the disorder. The classification of individuals as psychopaths then becomes less hostage to psychiatric prejudices about morally deviant behavior and more determinable by evidence that can stand apart from such deviance. Accordingly, the evaluation of these individuals as mentally disordered becomes less controversial, perhaps even uncontroversial, and the issue narrows to whether this disorder, when severe, excuses the sufferer from moral responsibility for the wrongs that result from it.

At the same time, those who press this strengthened argument for excusing psychopaths from moral responsibility rely on a characterization of their personality whose coherence is open to question. Psychopaths, they tell us, suffer from an affective disorder rather than a cognitive or volitional one. Jeffrie Murphy writes, for instance, "Unlike the psychotic, the psychopath seems to suffer from no obvious cognitive or volitional impairments" and three paragraphs later he continues, "Though psychopaths know, in some sense, what it means to wrong people, to act immorally, this kind of judgment has for them no motivational component at all.... They feel no guilt, regret, shame, or remorse (though they may superficially fake these feelings) when they have engaged in harmful conduct."[3] To be sure, Murphy hedges these remarks in ways that avoid commitment to an incoherent thesis. He says only that psychopaths seem not to suffer from any obvious cognitive or volitional impairment and only that there is some sense in which they have knowledge of right and wrong but are unmotivated by that knowledge. Still, in talking of moral feeling and motivation as if they were separable from the cognitive and volitional capacities that being a moral agent entails, he suggests a possibility that should not go unchallenged. Insusceptibility to moral feeling and motivation may imply cognitive impairments. And while tightening the category of psychopath by making certain emotional deficits essential to it naturally leads to identifying it as a type of affective disorder, it does not follow from this identification that the condition entails no cognitive disorder. In particular, it does not follow that it entails no deficiency in the faculties of moral judgment.

The question is an old one. It is a question of internalism in one sense of that now protean term. Does a moral judgment, specifically a judgment of what it is right to do or what one ought to do, imply a motive to moral conduct or a liability to moral feeling? Could moral judgments be truly practical or action-guiding if making them did not have this implication? The question is at the heart of many of the disputes in metaethics that have dominated Anglo-American moral philosophy in this century. Long running disputes about the concept of a moral judgment, about the meaning of the words 'right' and 'ought' when they are used to express moral judgments, and about the status of the properties, if any, that one predicates of actions when one makes such judgments are in large part due to intractable disagreements on this question. Consequently, if we treat the question from within metaethics, we cannot expect to get very far toward determining whether being a psychopath entails some deficiency in one's faculties of moral judgment. Whether the judgments of right and wrong psychopaths make are genuinely moral judgments, whether they use the words 'right' and 'ought' in a distinctively moral sense, and whether they predicate moral properties of actions when they say that an act is right to do or ought to be done are no easier to determine than the same questions asked about ordinary moral agents on those occasions when they judge that it would be right to do some act but are unmoved to do it and feel no regret or remorse about failing to do it. Naturalists, descriptivists, and realists will give us one set of answers. Emotivists, prescriptivists, and projectivists will give us another. And nothing in the behavior or thought of a psychopath, as distinct from that of an ordinary moral agent, argues for one set as against the other.

We can avoid the impasses of metaethics, however, if we treat the question from within psychology. The psychopath makes judgments about what it is right to do and what a person ought to do, in some sense of 'right' and in some sense of 'ought', yet is insusceptible to moral feeling and moral motivation. Accordingly, rather than ask whether this insusceptibility implies an inability to make moral judgments, we should consider different cognitive operations that can yield judgments about what it is right to do or what one ought to do and ask whether the insusceptibility implies an inability to engage in one or another of these. We can then leave for metaethics the question of whether any of them is essential to making moral judgments. Hence, the question of internalism, as a question asked within psychology, is a question about cognitive involvement in moral feeling and moral motivation. To be more exact, it is a question of whether some cognitive operations that yield judgments of what it is right to do and what one ought to do also yield motives to act in accordance with those judgments and liabilities to feelings of shame, regret, or remorse over failures to heed them.

We have now reached the principal question of our study. It has, as I have formulated it, two parts, one about motives and the other about feelings. They do not, however, require separate answers. Given that a person could not be susceptible to these feelings if he were insusceptible to the motives and conversely, answering either part suffices for answering the other. We can thus simplify the study by limiting it to one of the parts, and since the question of internalism with respect to motives is more tractable, we will make steadier progress by pursuing it and leaving the question with respect to feelings aside.

The first thing, however, that we must note, in focusing on motives, is the distinction between the internalism of hypothetical imperatives and that of categorical ones.[4] The distinction corresponds to two different ways in which a judgment about what it is right to do or what one ought to do could have motivational force. On the one hand, it could have such force relative to a desire the stimulation of which is independent of the cognitive operation that yields the judgment. This occurs where the operation is means-to-ends reasoning and the end being reasoned about is the object of a desire that is the occasion of the reasoning. On the other hand, the judgment could have motivational force in itself. This would occur if the force originated in the same cognitive operation that yielded the judgment. The first case is that of hypothetical imperatives; the second is that of categorical. Clearly, some cognitive failure would have to explain insusceptibility to the motivational force of the latter, whereas it need not explain insusceptibility to the motivational force of the former since such insusceptibility could be explained, instead, by the absence of the desire relative to which the judgment is made. Consequently, only the internalism of categorical imperatives gives a definite answer to our question about psychopathy. This means that, in considering cognitive operations that yield judgments about what it is right to do or what one ought to do of which internalism is true, we will need to attend to the kind of internalism it is. Ultimately, what we are after in treating the question of

internalism from within psychology is whether the internalism of categorical imperatives is true of any of these judgments.

To treat the question of internalism in this way presupposes that a
person can possess knowledge of right and wrong in different forms that
correspond to different levels of sophistication or maturity if not incompa-
rably distinct types of thought. At the least sophisticated level is mere
knowledge of the conventional moral standards observed in the person's
community, knowledge that one normally acquires fairly early in one's
upbringing. Later, as one's relations with others become more complex,
one comes to understand the reasons for these conventions and to comprehend the ideals that give them meaning. One thus acquires more sophisticated knowledge. Psychopaths presumably, being minimally socialized, at least have knowledge of their community's conventional moral
standards. It follows then, given the tighter construction of psychopathy
we are now using, that a person could possess such knowledge without
being motivated to act on it. To what degree psychopaths possess the
more sophisticated knowledge that comes from understanding the reasons
for the conventions and the ideals that give them meaning is, however,
uncertain. Consequently, whether a person could possess knowledge of
right and wrong at increasingly deeper levels of sophistication without
limit and still be insusceptible to moral motivation is a different matter.
The question of internalism, even if closed at the shallowest levels, is open
at deeper ones.[5]

Kant's ethics offers, arguably, the most important modern account of
this deeper knowledge of right and wrong. On his account, such knowledge consists of a formal principle that governs the workings of practical
reason. The statement of the principle that most clearly displays its formal
character is its first formulation in Kant's system. Loosely put, this first
formulation is that one act only on those rules that one can understand and
endorse as rules for all rational beings or, as Kant would say, universal
laws. Fundamental moral knowledge then, on this account, is knowledge of
a formal criterion of validity, the criterion of universalizability, that one
applies to rules of action, including conventional moral standards, but also
including personal imperatives, institutional directives, and other more
immediate practical principles. Indeed, Kant thought that in every rational
action there was at least implicitly some rule the agent followed, and in
this way he brought all rational conduct within the scope of his fundamental principle of morality.[6] It will be useful to adopt Kant's view of rational
action since doing so will simplify our discussion without affecting the
argument. Let us assume, then, as part of this account, that every rational
action proceeds from some rule. Accordingly, our deeper judgments of
right and wrong in every case consist in applying the criterion of universalizability to this rule.

How one is supposed to apply the criterion, however, is a controversial matter. There is no agreement either on what method of application Kant intended or on what method makes the most sense. Nonetheless, the point of applying the criterion is clear. It is to enforce consistency in practical thought and moral judgment. For if I am required to understand the rule of my action as a universal law and endorse it as such, then I cannot decide to do something that I would object to if it were done by another whose circumstances were relevantly similar to mine. Likewise, I cannot excuse myself or excuse a friend, for that matter, from doing an action I think others ought to do if our circumstances are relevantly similar to theirs. Understanding and endorsing the rule of one's action as a rule for all rational beings means applying it, without exception, to anyone whose situation meets the description contained in the rule of the circumstances that trigger its application.

The criterion thus serves as a filter on a person's reasons for action since rules of action specify such reasons. How it filters these reasons is not hard to see. If a person in considering whether to do an action to which he is inclined—double-parking, say, when in a hurry—tests its rule by applying the criterion of universalizability and finds that it fails, he must then, on penalty of being inconsistent, abandon further consideration of acting on that rule. This means he must dismiss as reasons for double-parking those facts about his circumstances that he would have taken as reasons if he had accepted the rule without testing it or had mistakenly thought it had passed. In short, a person who applies the criterion of universalizability in considering whether to double-park will exclude from being a reason for the action any fact that he would reject as a reason for another's double-parking if the latter's circumstances were relevantly similar to his own. He will conclude, then, unless of course he finds a reason that survives the test of universalizability, that he ought not to double-park.

Furthermore, his excluding from being reasons for double-parking facts that fail the test should correspond to his checking the inclination to double-park that initiated his deliberations, for otherwise the test would be idle. An aversion to inconsistency, it would seem, thus combats and, if sufficiently strong, suppresses the inclination to double-park when reasons for the action are invalidated by the test of universalizability. Application of the criterion would appear, then, in this case to yield both the judgment that one ought not to double-park and a motive to abstain from such action. It would appear, in other words, to be a cognitive operation that can at once yield judgments of what one ought to do and motives to act in accordance with those judgments. And if this is indeed so, then it shows that one can draw from Kant's ethics an account of our deeper knowledge of right and wrong that supports taking internalism to be true of that knowledge.

Of course, if a person's aversion to inconsistency is due to a desire to be rational that he acquires in the same way as he acquires the desire to be financially secure, then the motive that applying the criterion of universalizability yields is ultimately external to the cognitive operation. In this event, the account would at most support the internalism of hypothetical imperatives. It would be similar, that is, to an account of our knowledge of annuities that showed how one's calculations of the interest earned from an annuity can at once yield judgments about accrued benefits and motives to purchase annuities that would have those benefits. But if one holds, as Kant did, that an aversion to inconsistency is inherent in reason, that a rational mind spontaneously seeks consistency, then the motive would be internal to the cognitive operation and the account would therefore support taking the internalism of categorical imperatives to be true of our deeper knowledge of right and wrong.

The statement that an aversion to inconsistency is inherent in reason is open to interpretation, however. There is more than one thesis it could express. The least controversial is that the mind abhors contradiction. No doubt this thesis too has its share of opponents. Someone somewhere has surely argued that even the Law of Contradiction is a social construction, part of the knowledge/power regime that oppresses us all. I suspect, though, that the thesis enjoys widespread acceptance among philosophers and psychologists, acceptance that is by no means confined to enthusiasts for computer models of the mind. Be this as it may, the thesis that the mind abhors contradiction is too weak to justify taking the motive that applying the criterion of universalizability yields to be internal to that cognitive operation.

This is because basing a decision to act on reasons that fail the test of universalizability does not entail a contradiction The inconsistency it entails is of a different sort. If I decide to double-park, for instance, because I'm in a hurry, though I realize that I would object to another's double-parking even if he too were in a hurry and faced circumstances that weren't relevantly different from mine, then it's likely that I would manufacture an excuse to ease my conscience. But suppose I didn't. Suppose, instead, I thought to myself, while setting the parking brake and realizing that I would object to another's doing what I was doing, "I may double-park here; after all, I'm in a hurry and who's going to stop me!" In this case—call it the case of the psychopathic thought—I would in effect be drawing conclusions about what I may do and what another ought not to do that could not both be justified. Indeed, my conclusions would be arbitrary, since I would be treating my circumstances as justifying a license to double-park but denying that another's, which are not relevantly different from mine, provide similar justification. But to say that these conclusions are arbitrary is to say that there is a lack of uniformity or regularity in my drawing them, not that there is a contradiction between them. Inconsistency in the sense of lack of uniformity is not the same thing as inconsistency in the sense of contradiction.

Still there appears to be something illogical about my drawing these conclusions. After all, if I think that I may double-park but that another, whose circumstances are relevantly similar to mine, ought not to, then either I am ignoring facts about my circumstances that, as facts about his, I take as reasons against his double-parking or I am ignoring facts about his circumstances that, as facts about mine, I take as reasons for allowing me to double-park. Yet nothing in my circumstances as compared with his appears to justify this differential treatment of similar facts. The arbitrariness of my drawing these two conclusions qualifies as illogical, it would appear, in virtue of being a violation of the Principle of Sufficient Reason.[7] Perhaps, then, one could interpret the statement about the mind's inherent aversion to inconsistency as the thesis that the mind abhors such arbitrariness as well as contradiction. Obviously, this thesis would warm the heart of any good rationalist. But it would also leave many a voluntarist cold.

The difficulty in taking the Principle of Sufficient Reason as a principle of logic from whose violation every rational mind shrinks, the voluntarist would argue, is that not every rational mind feels compelled to justify decisions and judgments he or she makes that are to some degree arbitrary. Suppose on Wednesday I decline your offer of cream to put in my coffee, having accepted a similar offer from you the day before. My behavior might seem odd in the absence of any relevant difference between Tuesday's and Wednesday's circumstances, but its oddity might not bother me in the least. To be sure, some people have a low threshold of tolerance for inconstancy in their lives and in the absence of a reason justifying a reversal of decisions would feel uncomfortable about declining Wednesday's offer, having accepted Tuesday's. But there are others who have no qualms about seeing today's decision as independent of yesterday's and thus as needing no justification despite its reversal of yesterday's, and I may be one of them. The important thing is that their greater tolerance of inconstancy does not impugn their rationality. An aversion to the kind of arbitrariness to which violations of the Principle of Sufficient Reason belong is therefore external to the cognitive operation in which the Principle acts as a constraint on one's practical decisions and judgments. Or so the voluntarist would argue.

At this point, however, it is necessary to distinguish, at least in the case of the psychopathic thought, between the decision I reach and the judgments I make. For the voluntarist might be right to deny that the Principle of Sufficient Reason must constrain the former and wrong to deny that it must constrain the latter. The difference between the two is that the former is detachable from the reasons on which it is based and the latter are not. The former is detachable in that decisions, resolutions, and other acts of mind that, as we say, commit the will go beyond the reasons one has for such commitment. Those reasons may in effect dictate that the action be done, yet indecision or irresolution may keep one from deciding to do it. Conversely, one may decide to do something on the spur of the moment, as a matter of whim, or as a sheer act of will, which is to say for no reason at all. "I'll do it" is the natural expression of a practical decision, and this assertion is intelligible independently of there being reasons behind it. By contrast, judgments about what one ought or ought not to do are not detachable in this way from the reasons on which they are based. The judgment that one ought to do a certain action implies that one's circumstances require or merit the action, and therefore it implies that certain facts about those circumstances are reasons for doing it. The judgment, in other words, is not an intelligible thought independently of there being reasons behind it. One cannot coherently think, "I ought to do such and such, though there is no reason for my doing it."

Applying these points to the case of the psychopathic thought, then, we can say that, while I might not be constrained by the Principle of Sufficient Reason to decide against double-parking, since the decision goes beyond whatever reasons might dictate that I not double-park, the principle should constrain me to conclude that I ought not to double-park, given that I think that another, whose circumstances are relevantly similar to mine, ought not to. For if his circumstances require that he not double-park and they are relevantly similar to mine, then by the Principle of Sufficient Reason mine must also require that I not double-park, and because the conclusion that follows from my understanding of my circumstances in comparison with his cannot be detached from the reasons on which it is based, I am not free to ignore the Principle. Hence, when I conclude to the contrary that I may double-park, I have not merely shown a greater tolerance of inconstancy than a person who feels constrained to apply the same considerations to himself as he applies to others, I have actually shown muddled thinking. The absence of any aversion to the arbitrariness of my permissive judgment, in other words, implies a cognitive failure. The voluntarist, in this case, in denying that the Principle constrains judgments about what actions my circumstances require or permit, misapprehends the nature of those judgments.

It would be hasty, though, to infer from this result that internalism is true of our deeper knowledge of right and wrong as it is represented by Kant's initial account of morality's fundamental principle. The problem is that to see one's circumstances as relevantly similar to another's circumstances is already to be sensitive to the practical consequences of the comparison, for one cannot know which similarities are relevant and which differences are irrelevant without knowing what they are relevant and irrelevant to. Consequently, if one is unprepared to regard the interests of others as worthy of the same consideration as one's own and therefore unprepared to accept the practical consequences of such a fair-minded outlook, one may not see as relevant similarities that a person who is so prepared does see as relevant and one may see as relevant differences that that person sees as irrelevant. An unwillingness to suspend or revise the pursuit of one's interests, even where persistence in that pursuit harms others, may thus show itself in one's regarding as irrelevant similarities between one's circumstances and those of another that a more fair-minded person would regard as relevant. It may thus show itself in conflicting practical judgments like those in the case of the psychopathic thought without one's thereby running afoul of the Principle of Sufficient Reason. The Principle, in other words, could be a potent constraint on one's practical thought yet never actually constrain that thought in a way that confirms internalism.

The problem, clearly, arises from there being different ways in which people view themselves in relation to others. The egocentric agent, the agent who sees the effect of his actions on other people's lives as much less important than their effect on his own (and conversely), will not regard his circumstances as relevantly similar to other people's when his present him with opportunities for improving his life, advancing his interests, etc. and theirs present them with the same opportunities for improving their lives, advancing their interests, etc. Hence, his applying the test of universalizability need never yield a motive for action that opposes his self-interested motives since it need never transform his opposition to another's self-interested action—particularly one that adversely affects his interests—into opposition to his similarly self-interested actions. It need never transform his complaints about another's double-parking when it blocks his path into self-criticism of his own plan to double-park, even when he realizes that his double-parking will block the paths of those behind him. Of course, when a fair-minded agent applies the test in circumstances like those we are imagining, the application will yield a motive for action that opposes the agent's self-interested inclination to double-park. But—and here I appeal to Mill's methods, specifically the method of difference—we cannot infer that the motive it yields is internal to the cognitive operation that produced it, for this motive may originate in desires that a fair-minded agent has acquired and an egocentric agent has not, in which case it would be external to the operation when the fair-minded agent carried it out and nonexistent when the egocentric agent carried it out.

Perhaps, though, the way the egocentric agent views himself in relation to others implies impaired cognition. Perhaps, the egocentric view itself, treated as a kind of policy its agent adopts, fails the test of universalizability.

The most carefully constructed and fully developed argument for this latter thesis appears in Alan Gewirth's widely discussed book *Reason and Morality*.[8] The argument has received such thorough criticism that any further examination will now seem like an unnecessary autopsy. So I can afford to be brief. At the same time, much of the criticism has focused on Gewirth's claim to have derived fundamental moral principles from the bare idea of rational agency, and this claim, because it raises the questions of metaethics we have shelved, is beyond the scope of our study. So in shortening Gewirth's argument to its case against the rationality of the egocentric view and fixing it in relation to our study, I'll be looking at a less popular target of criticism and from a new angle. This should prove instructive.

Gewirth's strategy, as I've indicated, is to consider individuals as rational agents in abstraction from their particular aims and desires and their particular circumstances. His argument, in a nutshell, is this. Consider a rational agent, Molly Malloy, to give her a name. Qua rational agent, Ms. Malloy has purposes, which she necessarily thinks are good because, being her purposes, she cares about their fulfillment. Accordingly, she also regards her freedom and well-being as good since being free and tolerably well are necessary to her fulfilling her purposes, which, as we noted, she thinks are good. It follows that she will oppose interference by others with her freedom and well-being since such interference deprives her of what she needs as a rational agent. She will thus think, again in abstraction from time, place, and other particulars, "Others ought not to interfere with my freedom and well-being." Since her sole reason for this thought is that she is a rational agent, an individual with purposes that are good, she must then acknowledge, on penalty of violating the Principle of Sufficient Reason, that she ought not to interfere with the freedom and well-being of others. They too are rational agents, individuals with purposes that they regard as good, which are facts she cannot deny. Hence, the Principle constrains her to adopt a fair-minded view of her relations with others.

Having abstracted away all of the particulars of the agent's circumstances, Gewirth thinks there is now nothing to distinguish one rational agent from another, and hence whatever forbearances an agent demands from others he must accept as forbearances that others would be justified in demanding from him. There is no personal peculiarity, Gewirth would say, that Ms. Malloy could invoke as a relevant difference between herself and others. There is only the fact of her rational agency, which she has in common with others and which is the sole fact on which she bases her opposition to interference by others with her freedom and well-being.

Closer inspection of Gewirth's argument, however, reveals a mistake in this last point. Ms. Malloy does not base her opposition solely on the fact of her rational agency. She also bases it on her judgment that her purposes are good, which is to say, on her interest in fulfilling those purposes. Since she need not have any interest in whether others fulfill their purposes, she need not judge that their purposes are good. She may even judge that they are worthless or evil. Hence, she can invoke a relevant difference between herself and others, for as she views herself in relation to them, she sees that she acts for purposes that are good, and if no one else, as far as she can tell, is acting to advance her purposes, then she may also see that no one else is acting for purposes that are good. Hence, she may, without running afoul of the Principle of Sufficient Reason, think that others ought not to interfere with her freedom and well-being and at the same time think she may interfere with theirs. As she sees things, her freedom and well-being are goods, necessary goods as Gewirth would say, but theirs are not.[9] You might wonder, I suppose, whether this description of her is too fantastic, whether there really could be someone who saw value only in her own purposes and in what contributed to their fulfillment. But remember that the ultimate object of our inquiry is the psychopathic mind, and as a description of a psychopath's view of himself in relation to others this one shouldn't seem all that far-fetched.[10]

The description does contain one curious feature, though, a riddle, if you will, that arises when one asks, "How can Ms. Malloy have the thought that others ought not to interfere with her freedom and well-being? What could be its basis?" She can't be thinking, for instance, that others are forbidden by law from interfering with her freedom and well-being, or that morality forbids it, or that some other norms of proper behavior to which others are subject proscribe such interference. Laws, principles of morality, and other norms of propriety are not available to her merely as a rational agent, at least not at this early stage of Gewirth's argument. Nor can she be thinking that others would be well-advised not to interfere with her freedom and well-being, that a policy of non-interference would be to their advantage. Her thought here cannot be that of someone realizing, say, how angry she gets when others interfere with her and worrying about the consequences for them of her belligerence. It is not comparable to the thought "others ought to stay clear of me" that someone might have who posed a threat to others, owing, say, to a contagious disease. For as someone who cares only about fulfilling her own purposes, Ms. Malloy has no interest in other people's welfare and therefore nothing in her orientation toward others supports her thinking about what would benefit them. There are of course interpretations of her thought on which it is not a directive or imperative, but none of these is possible in the context of Gewirth's argument, since on any the question of whether Gewirth had derived a principle of right and wrong would not even arise.

To some philosophers, the answer to this riddle is simple. The description of Ms. Malloy I have given is incoherent. Since her thought that others ought not to interfere with her freedom and well-being must be an imperative based on reasons they have to refrain from interfering, she must be using 'ought' either categorically, in virtue of some authoritative rule to which others are subject, or hypothetically, in virtue of some advantage that others would gain from doing what she thinks they ought to do. That is, she must be assuming that they have either principled reasons not to interfere or reasons based on their interests and desires if she is using it intelligibly.[11] Yet neither of these alternatives makes sense, given that nothing more is true of her than that she is a rational agent, an individual with purposes that she regards as good, and that she acknowledges the existence of other rational agents, other individuals with purposes that they regard as good. The upshot of this answer, then, is that no judgment about what others ought to do can issue from a perspective like hers, the perspective of mere rational agency with which Gewirth's argument starts. His argument goes wrong in that it assumes that the mere rational agent can make such judgments, judgments to which the criterion of universalizability then applies, and my description simply inherits the same error. Thus, according to this answer, the opposition between Gewirth and me over what must follow from applying the criterion of universalizability to Ms. Malloy's judgment about what others ought to do is a mirage. No genuine opposition can arise from applying the criterion to an unintelligible judgment.

The answer is certainly congenial to the immediate point I am making. For it makes no difference to my rejection of Gewirth's argument whether the argument fails for the reason I gave or for the reason this answer furnishes. Still, I think the answer is mistaken, and the mistake does matter to the larger question at hand. There are egocentric uses of 'ought' that a less philosophically driven account of acceptable English should recognize as intelligible.[12] Married life provides familiar examples. Sometimes, when my wife and I are eating out and she is torn between two desserts, she will decide on one and tell me that I ought to order the other. While admittedly her instruction to me is open to various interpretations, an obvious one is that she thinks I ought to order the second dessert because she wants to taste it. At such times she is thinking of me as a helpmate rather than an independent agent, and consequently my own desires and interests do not figure in her calculations. She is not thinking, that is, of what I would like, imagining, for instance, which of the desserts would be tempting to me. Rather she is thinking of what I can do to help her get what she would like. Her use of 'ought', in this case, is egocentric, but intelligible nevertheless. As she views my circumstances in relation to hers, they merit my ordering the dessert because she desires it.

This sort of case is intelligible because we understand how two people who have shared a life for many years can come, in some situations, to lose sight of the boundaries that separate them as autonomous agents and to regard each other as extensions of themselves. Suppose, then, that Ms. Malloy viewed others in a similar way. Suppose that, though she recognized others as rational agents, individuals who had purposes that they regarded as good, she nonetheless regarded them as instruments of her will. "They may regard their purposes as good," she might say, "but it is my purposes, and not theirs, that truly matter. Hence," she would conclude, "they ought not to interfere with my freedom and well-being." And this conclusion, given her view of others as instruments of her will, would be perfectly intelligible.

Of course, if she were to imagine having the purposes that another person has, then she would see how purposes other than her own can matter too. That is, if she regarded these purposes as this other person would, she would then see that they were worth pursuing and accordingly conclude that she and others ought no more to interfere with his freedom and well-being than he and they ought to interfere with hers. But to arrive at this conclusion requires something besides applying the criterion of universalizability to her own judgment that others ought not to interfere with her freedom and well-being. It requires instead empathy with the person with whose freedom and well-being she judges that others ought not to interfere. Moreover, the empathy it requires must involve not only taking this other person's perspective and imagining the feelings of frustration or anger, say, that he would feel as a result of being interfered with but also understanding his purposes as generating (in conjunction with his circumstances) reasons for action even as one realizes that these purposes and reasons are independent of one's own. Only if this latter condition is satisfied can we say that Ms. Malloy recognizes the other person as a separate, autonomous agent. Only then can we say that she has advanced beyond the egocentric view.

To define the empathy capacity for which implies that one has advanced beyond the egocentric view, it is necessary to distinguish it from emotional identification.[13] Both involve one's taking another's perspective and imaginatively participating in this other person's life. But it is distinctive of empathy that it entails imaginative participation in the other's life without forgetting oneself.[14] The same is not true of emotional identification. Indeed, when such identification is strong and one's own identity weak or budding, the result is likely to be a loss of the sense of oneself as separate from the person with whom one identifies. Thus a boy who so strongly identifies with a favorite ballplayer that every game is an occasion for intense, vicarious ball playing does not manifest an empathic understanding of his hero. Rather he makes believe that he is this player and loses himself in the process. He takes the latter's perspective and imaginatively participates in the player's trials, successes, and failures, but in doing so he may merely be transferring his own egocentricity from one perspective to another. To empathize with another, by contrast, one must recognize him as separate from oneself, a distinct person with a mind of his own, and such recognition requires that one retain a sense of oneself even as one takes up the other's perspective and imaginatively participates in his life.

At the same time, empathy must involve more than seeing another as a separate site of mental activity and imaginatively participating in that activity if it is to imply an advance beyond the egocentric view. Yet making its involving more a defining condition of empathy would mean that the emergence of empathy in children as they grow older was a later and more abrupt development than it has seemed to those who have studied the phenomena.[15] Consequently, to capture its gradual emergence from early experiences of shared feeling, it is necessary to conceive of empathy instead as taking increasingly mature forms as one's understanding of what it is to be a human being and to live a human life deepens. Accordingly, one's egocentricity will be presumed to begin to weaken at some point as one's capacity for empathy matures, and advancing beyond the egocentric view would then be a matter of developing and exercising more mature capacities for empathy.

What argues for this developmental conception is the general outline of how empathy in children emerges and matures on which recent accounts by developmental psychologists, though they differ on specifics, agree.[16] Children, on these accounts, are capable, even at a young age, of responding in kind or solicitously to the behavior of others that signals their feelings and emotions, and in these responses one can see evidence of the child's early capacities for empathy. Young children, that is, indicate through their behavior—toward companions, for example—that they both share their companions' feelings and recognize that those feelings are distinct from their own. Yet a young child is far from seeing its companion as an autonomous agent or imaginatively participating in any more of its companion's life than the events with which it is immediately confronted. Its empathy is thus restricted to the immediate feelings, sensations, and emotions that another is experiencing, a restriction that reflects its limited comprehension of human life. This restriction then eases as the child learns to see people as having an existence that goes beyond the immediate situation in which it finds them. Correspondingly, it becomes capable of a maturer empathy in which the recognition of others as architects and builders, so to speak, of human lives is more pronounced. In taking another's perspective, it sees the purposes that give extension and structure to the other's life and sees those purposes as worthwhile, as purposes that matter. In this way it comes to recognize others as autonomous agents and to participate imaginatively in their separate lives.

Having a developmental conception of empathy is, needless to say, of signal importance to assessing the role of empathy in the deeper knowledge of right and wrong that people, as they mature, normally acquire. Thus, while reflection on characters such as Molly Malloy leads to seeing empathy as having a critical role in this knowledge, it would plainly be a mistake to infer that empathy *simpliciter* had this role. For the empathic responsiveness of young children to the feelings and emotions of their companions, though it certainly shows the potential for their acquiring deeper knowledge of right and wrong—indeed, it is hard to imagine a program of moral education that did not center on its cultivation and training—is nonetheless consistent with their having as egocentric a view as Ms. Malloy's. What appears to be critical, in other words, is empathy in a maturer form.

The same point applies to the troublesome case of sadistic pleasure. The case is troublesome because the pleasure a sadist gets from, say, assaulting someone is typically increased by his imagining his victim's pain. The sadist thus exhibits empathy inasmuch as he shows that he is taking in his victim's suffering, imagining, say, its course and intensity. Yet the sadist's empathy does not count against his having an egocentric view. The reason is that his empathy, like that of a young child, does not imply a recognition of his victim as an autonomous agent. The sadist, one might say, in getting pleasure from another's pain, fails to take in the whole person. He revels in the pain and suffering he has produced in that person but does not see beyond these particular feelings and emotions to the life his victim is living or the purposes that give it extension and structure. He does not see those purposes as worthwhile, as purposes that matter. How could he, one wants to ask, and still get pleasure from their frustration? He can, to be sure, like Molly Malloy, acknowledge that his victim is an autonomous agent, someone who has purposes and who regards his purposes as worthwhile. But acknowledging that a person regards his purposes as worthwhile is hardly the same thing as seeing from that person's perspective that his purposes are worthwhile. The sadist's empathy, in other words, is not of a form that could be critical to our deeper knowledge of right and wrong. It is not of a mature form.

Suppose, then, that judgments of right and wrong implying this deeper knowledge entail empathy in its mature form and that anyone who has and exercises a capacity for mature empathy makes such judgments. Suppose, that is, a different account of the deeper knowledge of right and wrong one acquires as one comes to understand the reasons for the conventional moral standards of one's community and to comprehend the ideals that give them meaning, an account on which judgments about what it is right to do or what one ought to do follow from one's empathizing with another or others rather than from one's applying the test of universalizability. Would internalism be true of such knowledge on this account, and would it be the internalism of categorical imperatives? Would the account, in other words, give us reason to conclude that psychopaths suffer some cognitive disorder? If it did, then their disorder would be a deficiency in their capacity for empathy, though the deficiency need not be so severe as to render them entirely incapable of empathy. Indeed, given the undoubtedly large intersection of sadists and psychopaths, it could not be that severe. It should be severe enough, however, that it prevented psychopaths from seeing others as autonomous agents, severe enough, that is, that it kept them bound to an egocentric view of their relations to others.

Initially, it may seem as though internalism of some kind must be true of the knowledge of right and wrong that one would exercise in taking up other people's perspectives, seeing their purposes as purposes that matter, and then judging that such and such ought to be done. If Ms. Malloy, for example, as a result of taking another's perspective and thus coming to regard his purposes as purposes that mattered, judged that she ought not to interfere with his freedom and well-being, then her judgment would imply, so it seems, a motive to refrain from such interference, since to see someone's purposes as purposes that mattered is to be inclined to do what would help to accomplish them. In this case, the judgment would seem to follow directly from judgments about what the person whose perspective Ms. Malloy took would want her to do given his purposes and given that these purposes were the focus of her attention. Yet the judgment could not follow directly in this way unless Ms. Malloy had either forgotten her own purposes or reflexively subordinated them to those on which she was focused, and on either possibility one can infer that the judgment would not be the result of mature empathy. On either possibility, it would have to be the result of emotional identification.

The reason it would not be the result of mature empathy is that such empathy, while it brings one to see another's purposes as worthwhile, does not necessarily lead one to favor those purposes over other purposes —one's own, in particular—that one also regards as worthwhile. Thus, for Ms. Malloy's judgment to be the result of mature empathy, she would not only have to see the other's purposes as worthwhile but would also have to resolve and conflict between his purposes and hers in favor of his or allow that there was no conflict. Put generally, mature empathy has to be combined with sensitivity to the possibility of conflict among the different purposes one regards as worthwhile and a criterion for resolving whatever conflict among them may arise before it can result in judgments of what one ought to do or what it is right to do. And since many situations call for empathy with more than one person, the possibilities of conflict corresponding to interpersonal differences can be multiple. Any account of our deeper knowledge of right and wrong that made empathy in its maturer forms essential must, therefore, also include a sensitivity to the possibility of interpersonal conflicts and a criterion by which to arbitrate those conflicts. For this reason, it is a mistake to assume that, in exercising this knowledge, one concludes directly from seeing another's purposes as worthwhile and realizing what action would help the person to accomplish those purposes that one ought to do that action.[17]

The question, then, is whether internalism is true of the judgments about what one ought to do that exercising this knowledge yields. This is, moreover, a particularly problematic question when the internalism it concerns is the internalism of categorical imperatives. For the judgments are no longer merely judgments about what one ought to do in order to promote personal purposes, one's own or another's, that one sees as worthwhile. Rather they are judgments about what one ought to do as the best way of resolving interpersonal conflict among such purposes. And even on the assumption that these judgments imply some motive to act on them, it is not easy to determine from abstract consideration of the cognitive operations that would yield them whether this motive is internal to those operations or due, say, to an external commitment to an ideal of harmony in interpersonal affairs and relations.

At the same time, we might find support for ascribing such internalism to these judgments in a different place. Specifically, we might find that, as a fact of human psychology, to possess and exercise the capacity for mature empathy is to be disposed to accept some criterion for mediating harmoniously among the competing purposes of the many autonomous agents in whose lives one imaginatively participates. We might find, in other words, that, as a fact of human psychology, no one who was capable of taking the perspectives of different individuals and seeing seriatim that different, competing purposes corresponding to these different perspectives each mattered could rest content with the disharmony that this survey would yield or steel himself to the pull of all those purposes that were not compatible with his own. A psychopath, then, might suffer deficiency in empathic powers for either of two reasons: stunted development or regression to an egocentric view out of inability to tolerate the clash of different perspectives with which one in whom the capacity for empathy has matured is presented. Plainly, confirming or falsifying any of these hypotheses is beyond the methods of moral philosophy. It properly falls within those of developmental psychology and cognitive science. At this point, our philosophical study must look to these other disciplines for help in unraveling the mystery of the psychopathic personality.

218   John Deigh

Notes

I am grateful to Jonathan Adler for helpful comments several years ago on a paper from which this article descends. I have also benefited from conversations with Herbert Morris, Connie Rosati, and Ira Singer and from the comments I received from two editors of *Ethics* and from the audiences at Washington University's conference on ethics and cognitive science, Bowling Green State University, and the University of California, Riverside, where I presented an earlier draft of the article.

1. Truman Capote, *In Cold Blood* (New York: Random House, 1965); *Strangers on a Train* (Burbank, Calif.: Warner Brothers Pictures, 1951), directed by Alfred Hitchcock, screenplay by Raymond Chandler and Czenzi Ormonde.
2. See Hervey Cleckley, *The Mask of Sanity*, 5th ed. (St. Louis: Mosby, 1976), pp. 337–64.
3. Jeffrie Murphy, "Moral Death: A Kantian Essay on Psychopathy," *Ethics* 82 (1972): 284–98, esp. pp. 285–87.
4. See Thomas Nagel, *The Possibility of Altruism* (Oxford: Clarendon, 1970), pp. 7–12. The terms 'internalism of hypothetical imperatives' and 'internalism of categorical imperatives' are mine, not Nagel's, but they correspond to the distinction he draws between the internalism of Hobbes's ethics and the internalism of Kant's. See also my "Sidgwick on Ethical Judgment," in *Essays on Henry Sidgwick*, ed. Bart Schultz (Cambridge: Cambridge University Press, 1992), pp. 241–58.
5. Strictly speaking, the relative depth of the knowledge is irrelevant to whether the question of internalism with respect to it is open. For what keeps the question open is the possibility of there being some form of knowledge of right and wrong that no psychopath could possess, and the possibility of such knowledge does not rest on its level of sophistication as compared with the deepest level of the knowledge attributable to psychopaths. At the same time, making the supposition that any form of knowledge of right and wrong that psychopaths could not possess must be deeper than that which they do possess introduces a scheme that facilitates our treatment of the question of internalism, and it is for this reason that I make it.
6. See Immanuel Kant, *Groundwork of the Metaphysic of Morals*, trans. H. J. Paton (New York: Harper & Row, 1964), p. 80 (p. 412 in the Royal Prussian Academy ed.).
7. See Don Locke, "The Trivializability of Universalizability," *Philosophical Review* 78 (1968): 25–44.
8. Alan Gewirth, *Reason and Morality* (Chicago: University of Chicago Press, 1978), pp. 48–198.
9. Ibid., pp. 52–63.
10. See, e.g., Cleckley, *Mask*, pp. 346–48.
11. Statements of this view about the use of 'ought' in practical judgments are found in Philippa Foot, "Morality as a System of Hypothetical Imperatives," *Philosophical Review* 81 (1972): 305–16; and Gilbert Harman, "Moral Relativism Defended," *Philosophical Review* 84 (1975): 3–22.
12. Gewirth (p. 79) also argues for the intelligibility of this use of 'ought'. The argument's metaethical premises, though, invite the kind of dispute I mean to avoid.
13. I adapt here a point Max Scheler made in distinguishing among emotional contagion, fellow feeling, and emotional identification. See his *The Nature of Sympathy* (Hamden, Conn.: Shoe String, 1970), pp. 8–36.
14. The word 'empathy', it should be noted, does not have a settled meaning among those who write on the topic. The meaning chiefly varies between a kind of cognitive state and a kind of affective state, and I intend to capture the former in distinguishing it from emotional identification as I do. On the history and current variability of its usage, see Laura Wispé, "History of the Concept of Empathy" in *Empathy and Its Development*, ed.

Nancy Eisenberg and Janet Strayer (Cambridge: Cambridge University Press, 1987), pp. 17–37; and Janet Strayer, "Affective and Cognitive Perspectives on Empathy" in Eisenberg and Strayer, eds., *Empathy*, pp. 218–44.

15. For a useful survey of relevant studies, see Alvin I. Goldman, "Ethics and Cognitive Science," *Ethics* 103 (1993): 337–60.

16. R. A. Thompson, "Empathy and Emotional Understanding: The Early Development of Empathy," in Eisenberg and Strayer, eds., *Empathy*, pp. 119–45; and Martin Hoffman, "Interaction of Affect and Cognition in Empathy," in *Emotions, Cognition, and Behavior*, ed. Carroll Izard, Jerome Kagan, and Robert B. Zajonc (Cambridge: Cambridge University Press, 1984), pp. 103–31, and "The Contribution of Empathy to Justice and Moral Judgment," in Eisenberg and Strayer, eds., *Empathy*, pp. 47–80.

17. This point is well illustrated by the leading accounts of this sort. These are ideal observer accounts, such as R. M. Hare's, and contractualist accounts, such as T. M. Scanlon's. On either, judgments of what it is right to do or what one ought to do mediate conflicting interests and desires that reflect the competing purposes of the various people whose lives one must consider in making these judgments. While empathy is essential to considering these various lives, the judgments follow only after one applies the method of ideal observation or contractualist negotiation to resolve the conflicts among them and thus do not follow directly from judgments of what any of those whose lives one considers would want one to do given his purposes. See Hare, *Moral Thinking* (Oxford: Clarendon Press, 1981); and Scanlon, "Contractualism and Utilitarianism," in *Utilitarianism and Beyond*, ed. Amartya Sen and Bernard Williams (Cambridge: Cambridge University Press, 1982), pp. 103–28.

Chapter 12
Feeling Our Way toward Moral Objectivity
Naomi Scheman

However diversely philosophers of psychology have thought of emotions, they have mostly agreed on thinking of them as states of individuals. And most moral philosophers have agreed on thinking of emotions as for the most part inimical to the achievement of moral objectivity. I argue in this chapter that both views are wrong and that we can learn something about why they are by looking at how they are connected.

I

Arguing that emotions are not states of individuals can seem merely perverse. My first attempt to do so (Scheman 1983) has seemed so to most philosophical readers.[1] It is, of course, true that we ascribe emotions to individuals and that we speak of individual people as being in various emotional states. But there is a more precise and particular sense of what it is for something to be a state of an individual that I think can be made out clearly, and in this more precise sense, emotions (along with other complex mental "states," such as beliefs, attitudes, desires, intentions, and so on) are not states of individuals. The sense I mean is this: S is a (complex) state of an individual I only if the elements of S are related to each other in causal or other ways that make of S a complex entity independently of I's social context. (My concern will be to argue against this necessary condition; I have no interest in what might be sufficient conditions for something's being a state of an individual, or in any states that might be argued to be simple, such as qualia.)

The ontological distinction I am relying on is that between a complex entity and a jumble or a heap. Complex entities differ from jumbles or heaps in the relationships of coherence among their component parts. There is no way of drawing this distinction in any absolute way, since there will always be relationships to be found, or created, among any collection of entities. Rather, we need to ask whether some collection constitutes a complex object with respect to some particular theory or explanatory scheme (or some set of theories or schemes).

My argument will be that the coherence of emotions and other complex mental states is relative to irreducibly social, contextual explanatory schemes. In abstraction from particular social contexts, there is no theoretically explicable coherence among the behavior and the occurrent thoughts, feelings, and sensations (or whatever else one might take to be the components of an emotion; I have no commitment to this particular list, though it seems to get at what we do in fact point to when we identify emotions and other mental states). The distinction is akin to that between galaxies and constellations. Despite their both being made up of stars, galaxies are, and constellations are not, complex objects with respect to astronomy. The stars that make up a galaxy are related to each other causally and spatially, while the stars that make up constellations are related to each other only against the background of a set of stories about the night sky told by particular cultures on a particular planet; they are not even spatially contiguous except from the perspective of earth.

Emotions, I argue, are constellation-like, not galaxy-like. Their identity as complex entities is relative to explanatory schemes that rely on social meaning and interpretation. Emotions and other complex mental states differ from constellations in depending ontologically on explanatory schemes that have enormous depth and power and that we have every reason to continue to use—unlike astrology. The analogue of eliminative materialism would eliminate constellations from the furniture of the universe—not, I would think, inappropriately. But eliminative materialism as usually understood, which accepts my negative judgment on the possibility of even token identities between mental and physical states, would require that we jettison the explanatory resources of common sense, along with literature and the other arts, not to mention most of psychology, a prospect I find bleakly inhuman, as well as pointless.

It is, of course, true that many genuine states of individuals—diseases, for example—are identified and named only in particular social contexts, but the causal relationships that obtain among the causes and the symptoms are independent of those contexts; they are there to be discovered. There are intelligible questions to be asked about whether some particular pattern of symptoms constitutes a disease, and one important deciding factor is whether they hang together independent of their seeming to be somehow related, as, in the paradigm case, by being all caused by the same pathogen. If, on the other hand, the pattern of the symptoms is explicable in terms not of underlying somatic processes but of social salience, then it would be wrong to speak of a disease.[2] What is at issue is not what we call it, or even whether we call it anything, whether we find it interesting or important to name it, or whether its cause was or was not "biological" (pathogens can be social; to argue that a set of symptoms is caused by, for example, childhood abuse is not to argue that they do not constitute a disease). What is at issue is whether there is an "it" at all.[3]

This argument, drawn from Wittgenstein, is an explication of what he refers to as "the decisive move in the conjuring trick..., the very one we thought quite innocent," since all we have done is to say what it is we intend to look at, that is, "processes and states...[we] leave their nature undecided" (Wittgenstein 1953, sec. 308). What has, surreptitiously, been decided is that these supposed processes and states are complex entities according to some possible theory, yet to be developed. There must, by this picture, be some underlying mechanism that knits each process or state together, some structure for some future science to reveal. But why should this be so? What are we assuming, and what are we precluding, when we make this initial, apparently innocent move? What we are precluding is precisely what I want to suggest is the case: that emotions (and other complex mental "states") are situationally salient, socially meaningful patterns of thought, feelings, and behavior.[4] As such they may well be supervenient on physical facts, but only globally—that is, not only on physical facts in my body and not only on physical facts in the present or even in the past.

Consider, for example, these lines in Shakespeare's *Sonnet 116:* "Love is not love which alters when it alteration finds." There is something paradoxical in the claim—the same paradox as in Gertrude Stein's remark about Oakland: "There's no *there* there." In both cases there is something to point to, though the point of the pointing is to get one to see that there is, in a sense, no *thing* there, that by the standards by which we (more about "we" later) judge such things, this, whatever it might be, just does not count. The "it-ness" is not to be taken at face value: Shakespeare and Stein are urging us to have higher ontological standards, not to confer the dignity of coherence on what we ought, rather, to regard as a jumble or a heap. Jumbles and heaps can be pointed to; they can even be named but paradoxically: to do so is to adopt an interpretive perspective that one does not fully and wholeheartedly occupy. (In Sabina Lovibond's [1983] terms, it is to speak ironically, in inverted commas.) True love, or San Francisco, is "real" not because each has an inner coherence that false love and Oakland lack but because the patterns that lead us to see each as a coherent, complex object are salient from positions in which we wholeheartedly stand.

One might ask, in response to the denial that love that alters is (really) love: "If it isn't love, what is it?" I suggest that "it" might not be anything at all. Not that nothing is going on: rather, all that is going on—all the feeling, all the behavior—may not add up to anything, may not coalesce into a pattern we recognize as significant enough to name. It is as though one started out to choreograph a solo dance and midway through got bored and drifted off into aimless jiggling and sashaying. The dance, had one stuck with it, would have been a "something": in the absence of its completion, what we have is a jumble of movements, not ontologically distinct from the rest of what one is doing, before, after, and during that time. What I call the individualist assumption in the philosophy of psychology is the assumption that whatever we find it significant to name—whatever patterns of feeling and behavior we find salient enough to use in explanations of ourselves and others—must also pick out complex events, states, or processes with respect to some theory of the functioning of individual organisms (for example, neurophysiology). Only the strongest reductionist projects would support such an assumption, since it is precisely the assumption of type/type identities.

There are two things in particular to note about the account of love in Shakespeare's sonnet. One is that it is clearly intended to be normative; that is, we are being urged not to count as love anything that alters when it alteration finds. (And it may well be difficult not to see it as love, especially if the feelings are one's own; one may well feel injured or betrayed by whatever "alteration" seemed to destroy one's love and cling to the sense of the value of what one feels was stolen. Similarly if one is the object of the "love which is not love," one might cling to the idea that one was, really and truly, loved—so long as one was, say, young and beautiful.)

Second, as a normative account, it is potentially and unpredictably retroactive; that is, what seems to be love today might be judged not to be love because of something that happens tomorrow, something that need not in any sense be latent today. We are being urged to withdraw an honorific label from a set of feelings and behavior because of a failure of constancy in the face of change—not that it could not have been love if it failed that test, as though there were something else in which its authenticity consisted, of which constancy were a reliable indicator. Wittgenstein draws our attention to this sort of normativity: "Love is not a feeling. Love is put to the test, pain not. One does not say: 'That was not true pain, or it would not have gone off so quickly'" (Wittgenstein 1967, sec. 504). What is at stake is less what "it" is than what we are, what we care about, value, and honor, a refusal to be swayed by the intensity of passing feeling. And it makes sense to say this to us only if we are at risk of being so swayed.[5]

I want to consider further examples to try to make plausible the view of emotions as situationally salient, socially meaningful patterns of thought, feelings, and behavior, rather than as states that one might discover within particular individuals at a particular time. The point is that it is such patterns (not, for example, physiological causal connections) that make the components of an emotion hang together as a complex object rather than being a jumble or a heap. And, like pattern perception generally, the identification of emotions cannot be abstracted from the contexts in which the relevant patterns are salient.

The first example is drawn from the film *Torch Song Trilogy*, in particular, from the scene in which the central character, Arnold, is at the cemetery with his mother. Arnold is a gay man. He is, in fact, a drag queen by profession. He had a loving, committed relationship with another man, Alan, who was murdered by gay-bashing men in the street beneath their apartment. Arnold has buried Alan in the cemetery plot that his parents provided (presumably for the wife they hoped he would have). He goes to the cemetery to visit Alan's grave with his mother, who is visiting the grave of her husband, Arnold's father. Arnold's mother looks over at Arnold at his lover's grave and sees him reciting the Jewish prayer for the dead, and she asks him what he is doing. He says he is doing the same thing she is, and she replies in outrage that no, he is not: she is reciting Kaddish; he is blaspheming his religion. She finds it outrageous that he might take himself to be feeling grief, as she is feeling grief, and to have felt for Alan love, as she felt love for her husband.

That is, she looks at what Arnold is feeling in the context of Arnold's life and sees something quite different from what she felt in the context of her life. Sure, she thinks, he goes through these motions, says these words, even feels these pangs and aches, but they do not add up, hang together, or amount to the same thing as they would in the context of a proper marriage. (His feelings are no more grief than his saying those words is reciting Kaddish. To her they are equally meaningless jumbles.) Arnold, however, looking at his and her feelings, perceives the patterns differently from how she does; he weighs differently the similarities and differences, in the contexts of his and her lives, and ends up seeing the same feelings of love and of grief. It is the burden of the film, with respect in particular to a heterosexual audience, to persuade us to accept Arnold's mode of pattern perception and to reject as narrow-minded and heterosexist his mother's refusal to find in his actions and feelings the same valued coherence she finds in her own.[6]

A second example is drawn from popular songs, of which there are an enormous number, that couple love with potentially violent jealous possessiveness. In songs going back at least to those that were popular in the forties, one of the commonest themes, along with the "somebody done somebody wrong" songs, is the idea that love, in particular that which a man feels for a woman, must be reciprocated; she has no choice but to reciprocate it, and her turning her affections elsewhere is a possibility that, because what he feels is love, he cannot allow. Her simply failing to reciprocate his love may, in fact, constitute the "wrong" that he takes her to have done him.

Songs like this are interesting and important to think about, in particular because of the current controversy over the words in so-called gangsta rap. If we are going to take seriously the influence of popular song lyrics, we need to focus at least as closely on those that have filled the mainstream airwaves as on those that come from subordinated populations. What is it that we are doing, in particular, when we call the complex of sentiments in those popular songs love? We could, alternatively, pathologize this complex of feeling and attitude and behavior. We could call it something like "possession dementia" and enter it into the next edition of the *Diagnostic and Statistical Manual of Mental Disorders*. Or we could simply fail to see any "it" there at all. Valorizing and pathologizing are equally ways of bestowing coherence, however much we may believe that we are merely pointing to something—an emotion or a disease—that exists independent of our interest in it. The consequences of the elision of our interpretive activity in favor of a supposedly helpless recognition of "love" are, of course, disastrous, both because of the male behavior that is thereby licensed and because of the compliance it exacts from women who believe that such congeries *are* love, believe that it is something to be valued and honored.

On a socially constructionist account of emotions, as María Lugones argues (1987, 1991), we "make each other up," at least insofar as we set the parameters of intelligibility. We determine which patterns of feeling and behavior will count as love or grief or anger or whatever else. We also more personally and directly make each other up, as we help or hinder each other in seeing new or different patterns. For example, I have argued (Scheman 1980) that the central character in that paper, Alice, was helped by other members of her consciousness-raising group to see a pattern of depression, odd outbursts of impatience and pique, and a general feeling of undefinable malaise as anger. We can sharpen each other's pattern perception in the face of various kinds of resistance or fear.

Often, as is likely in this case, we will come to regard a particular pattern (Alice's anger) as so striking, once we come to see it, that a fully realist vocabulary is apt (she *discovered* she was angry). Speaking this way amounts to placing our interpretive activity in the background, taking it for granted, treating it as part of common sense. There is often, of course, good reason to do this, as others will have good reason to challenge it. What is important to note is that such challenges ought not to succeed if all they do is point out that Alice's anger is visible as such only if you stand *here* and look through *these* lenses. The response to that is, "Yes, of course, and what anger is any different? Do you have a problem with standing here, with looking through these lenses?" The answer to that question is likely to be yes: we cannot make judgments without standing on some ground or other, or see without lenses (even if only those in our eyes), but we need to be ready, when concretely challenged, to argue for the rightness of the choices we make, even those we may not have recognized as choices until someone questioned them.

For example, in the story I tell, the other women in the group spoke of their own anger, in the contexts of lives not unlike hers, and Alice's not regarding them as moral monsters or as wildly irrational helped, along with an evolving political consciousness, to create the background against which she could similarly see herself as angry. What goes on can be described as a kind of crystallization. And there is, of course, all the difference in the world between unacknowledged and acknowledged emotions. Once we acknowledge a significant pattern in our feelings and behavior, we are in a position to act differently. There is also all the difference in the world between the discovery of anger and its creation, especially its inappropriate creation, as some would argue that Alice was brainwashed by feminist antifamily propaganda.[7] I do not want to deny that there is such a difference, but there are no formal grounds—no grounds less contested than feminism itself—on which to base a judgment about which it is.

Conversely there is the possibility of justifiably chosen disavowal of feeling. For example, one can take as infatuation rather than as love feelings one judges to be inappropriate, feelings that, given how we think of ourselves as moral beings or as socially situated in the world, we think it would be better for us not to have. Or we could work toward overcoming prejudicial fears and antipathies by thinking that those feelings are incompatible with the people we take ourselves most fundamentally to be. We do not take ourselves simply not to have such feelings; rather, we redefine them. In the first case—redefining love as infatuation—we are likely to downplay the "it-ness" of what we are feeling, to fragment our responses, to attach them to various features of the situation, rather than letting them all cohere around the person we are trying to "get over." In the latter case—transforming judgment into phobia—we might rather focus on whatever coherence we can find not in the situation but in our history; we might choose to see as coherent, caused pathology what we might have been inclined to regard as situationally differentiated responses. Locating the explanatory coherence in ourselves, not in the objects of our judgment, is a central piece of identifying, for example, racism or homophobia.

How we do all of this shifting in pattern perception depends enormously on the interpretive resources, the encouragement or discouragement of those around us. Centrally, it depends on whom we take "we" to be—those with whom we share a form of life, whose responses we expect to be congruent with or touchstones for our own. Sympathy and empathy—feeling with and for each other—are therefore not just matters of response to independently given emotional realities. Rather, they change the context of experience and the context of interpretation. In so doing, they change the emotions themselves, frequently in ways that do not work by causing any changes in behavior or in immediate felt experience; such changes can, for example, be retroactive. A characteristic theme of Henry James's novels and novellas, for example, is the way in which subsequent events can utterly change the meaning of what someone felt and did—can, for example, make it true, as it was not before, that one person loved another. (I am thinking in particular of *The Golden Bowl, Wings of the Dove,* and "The Beast in the Jungle.")

II

The second argument is for the necessity of emotions to moral judgment. One of the reasons for this necessity has to do with the importance of moral perception to moral judgment—that is, much of the work of moral judgment is done in how we perceive a situation, what we see as morally relevant or problematic. Probably the moral theorist for whom such perception is most obviously central is Aristotle, for whom practical wisdom consists largely in this ability to perceive the salient features of a situation. But equally, if less obviously, for Kant, perceptions of salience are necessary to determine what the appropriate maxim is, that is, the one we need to test against the requirements of the categorical imperative: of all the ways I might describe what I am considering doing, which description best captures what might be morally problematic? (Do I ask if it is permissible to say something pleasing in order to make someone feel good, or do I ask if it is permissible to lie in order to make someone feel good?) Moral perception is arguably also important for utilitarians: everything I do potentially affects the pleasure and pain of myself and others, but not everything I do is appropriately subject to the felicific calculus. Any usable utilitarian moral theory is going to have to distinguish between actions that are and actions that are not morally problematic, and their having consequences for someone's pleasure and pain cannot be a sufficient ground. What is sufficient, however, presumably will have centrally to do with what those pleasures and pains are likely to be; in coming to an accurate description of what is morally salient about a situation, we need to be sensitive to the relation of our possible actions to the emotions of ourselves and others.

Emotions play various roles in moral epistemology, in helping us come to adequate moral perception. The first is that emotions partly constitute the subject matter of moral epistemology; that is, how I and others feel is often part of the morally relevant description of the situation. Furthermore, such descriptions are always given from the perspectives of emotional complicity. For reasons related to my first argument, these are not disinterested perceptions; they are, rather, the perceptions of people who are, at the very least, part of the background against which the emotions being described get the salience that they have.

The second role of emotions in moral epistemology is that of an epistemological resource. One of the clearest statements of this point is Alison Jaggar's (1989) discussion of what she calls "outlaw emotions": emotions that, according to the hegemonic view of the situation, one is not supposed to feel. Women, for example, are not *supposed* to be angered by whistles and catcalls as they walk down the street. We are supposed, rather, to feel flattered by them. One is not *supposed* to feel anger at certain kinds of jokes—sexist, racist, and so on—in certain social settings. Feelings such as these, which one is not supposed to have, are, Jaggar argues, often the best sign that there is something going on in the situation that is morally or politically problematic.

Outlaw emotions can work quite effectively against the phenomenon of gaslighting, a phenomenon that the poet and essayist Adrienne Rich (1979) has described, whereby we undermine our own and each other's perceptions. Something is going on that leads us to feel an outlaw emotion —for example, sexual attentions or sexual joking in the workplace that feels creepy and uncomfortable. We can be told and can learn to say to ourselves that we are being overly sensitive, prudish, or whatever else, and we thereby undercut the possibility of there being a critical edge to those emotions. A way of working against gaslighting is to take outlaw emotions seriously, as epistemic resources pointing us in the direction of important, often morally important, perceptions.

One way of putting this point is that unemotional perception is problematically partial, especially in its failure to include outlaw perceptions. The effect of ignoring or bracketing our emotional responses will frequently be that we gaslight ourselves into a failure to notice and learn from importantly critical perspectives, that we perceive only along the hegemonic lines of sight. (The problem is not confined to what we might think of as the context of discovery; it is not just that we are less likely to notice a wrong, but from a perspective uninformed by the relevant outlaw emotions, we will not acknowledge it even when it is pointed out to us. That is what gaslighting is about: a systematic biasing of the contexts of both discovery and justification.)

As Jaggar notes, not all emotional responses, not even all outlaw emotional responses, are valuable in this way. There is, I suspect, nothing to be learned from the annoyance some people feel in the face of having it pointed out to them that their behavior, including their choices of words, contributes to forms of social subordination. The defensibility of making such responses "outlaw"—that is, not how people are "supposed" to feel—rests in part on there being nothing to be learned from taking them seriously, nothing we do not already know about resistance to having one's privileges challenged. There may, of course, be much to learn from the fact that people do feel these ways, but the feelings are not themselves valuable epistemic lenses.

Much of the current dispute about "political correctness" can be seen as disputes about which responses ought appropriately to be seen as outlaw—not, that is, *outlawed*, if that even makes sense, but regarded as inappropriate, as epistemically unreliable. It follows from my first argument that any talk of emotions is deeply and pervasively normative, so there is no escaping such judgment of appropriateness, or any way of circumventing the politically tendentious nature of the judgments we make. Nothing simpler than attention to the specificities of a particular case will settle the question of what sort of epistemic weight we ought to put on a particular emotionally informed perception. Again, what is at stake is the identification of an appropriate "we" and our own, possibly problematic or partial, frequently critical, at times ironic, identification with it. Outlaw emotions can, for example, draw our attention to the dangers of excessive rigidity and a passion for purity, even with respect to values we wholeheartedly endorse. (Such was, of course, the original intent of the term "political correctness.")

III

I have been arguing that (1) any descriptive account of emotions is necessarily social, including normatively—that is, both generally and in specific cases, moral judgment enters into a descriptive account of our emotions—and (2) our emotional responses to situations and to each other necessarily inform the moral judgments that we make. I argue now that this circularity is benign and that it facilitates rather than undermines the possibility of the objectivity of moral judgment (and for that matter, the objectivity of the ascription of emotion, though it is moral judgment that I will focus on here).

In order to make this argument I need to explain what I mean by objectivity. I want to take as the motto for this account a quotation from Wittgenstein (1953, sec. 108): "The axis of reference of our examination must be rotated, but about the fixed point of our real need." What is our "real need" with respect to objectivity? I suggest, especially with respect to the objectivity of moral judgment, that the need is for moral knowledge that is "commonable," that is, stably sharable across a maximal diversity of perspectives. Can moral knowledge be objective in this sense? (Alternatively, one might ask the question: is moral *knowledge* [as opposed to moral opinion] possible?) I can see no reason why it should not be, but we need to consider the conditions of its possibility—that is, the conditions of the possibility of moral judgment that we have good reason to believe will be stable across a maximal diversity of perspectives, that will remain in place and not be upset when the situation is looked at from perspectives different from those that have informed it heretofore.

The usual problems with respect to objectivity—that is, the reasons people who believe in its possibility think that it is difficult to obtain—are conceived of as problems of partiality defined as bias. Seeing the problem in this way leads to such things as ideal observer theories or other theories couched in terms of desirable forms of ignorance. Factual knowledge may need to be maximized, but following the dominant epistemology of empiricism with respect in particular to emotions (which takes it as necessary insofar as possible to bracket them), the emphasis is on an epistemology of parsimony. We need to bracket off those influences that are taken to be problematic, where "problematic" means in practice "likely to produce diversity of opinion."

But we could see the problem rather as one of partiality defined not as bias but as incompleteness. That is, a judgment we have come to seems acceptable to us, but we cannot be sure that someone will not walk through the door and say, "That looks wrong to me"—and that in the face of that challenge we will either agree right off that their claim has standing against our consensus or be persuaded eventually to change our initial assessment of it or them as mad, ignorant, romantic, childish, or whatever else. If this is the way that we think of the problem, then what we need is not an epistemology of parsimony, guarding us in advance against the possibility of objections, but rather an epistemology of largesse.[8] We need to be expansive in who constitutes the "we" whose judgments and critical input help to shape the knowledge in question.

One way of addressing our "real need" with respect to objectivity is to ask what reason we have to believe that something we think we know will not be shaken by criticism coming from a yet unforeseen perspective. The best answer is that we have no reason to believe that. Our claim to objectivity rests precisely on the extent to which our practice makes such critique a real possibility. Rather than being able to argue in advance that no effective criticism could be forthcoming—because we have managed methodologically to anticipate and defuse it—we stake our claim to legitimacy, that is to objectivity, precisely on our record of having met actual, concrete criticism and thus earning the trust that we will do so in future. Objective judgments are more, rather than less, defeasible. What makes them count as knowledge—rather than, say, a working hypothesis—is their having come out of and been subjected to the critical workings of a sufficiently democratic epistemic community.

Largesse is not, however, the whole story. Not all voices are or should be equal. Not all perspectives afford usefully different vantage points. Not all criticisms should lead us to change our minds. The mistake of empiricist epistemologies—epistemologies of parsimony—has been to put faith in formal general principles of exclusion to erect a *cordon sanitaire* that would serve to discriminate a priori between those influences on belief that are appropriate and those that are not. But certainly in the realm of moral knowledge, as in any other knowledge that concerns human flourishing and harms, such faith is misguided. (Feminist philosophers of science and others have argued, persuasively, I believe, that it is misguided even in the "hard" sciences. See, for example, Longino 1990.) Discrimination—between the crackpot and the reasonable, the bigoted and the spiritedly partisan, the intolerant and the upright, the subordinated margins and the lunatic fringe—is both necessary and unavoidable. But nothing less substantive, less political, less controversial than fully messy engagement in the issues at stake, in the real differences between us, in the concrete ways in which various "we"s are constituted, will enable us to make these discriminations in ways that do not beg the very questions they are meant to decide. Emotions, in particular, can be invaluable epistemic resources as often as they can be biasing distractions, and there is no way of formally deciding which is the case.

I want to discuss briefly one particular role of emotions in achieving moral objectivity. The principal barrier to the actual achievement of objectivity is power, that is, the ways in which hegemonic social locations are constructed and others are marginalized, silenced, or distorted. The thinking of those who occupy positions of authority and privilege is problematically and systematically parochial. The knowledge claims that we produce, whether moral or factual, are in need of critical examination from the standpoints of the subordinated. The reason for such critique is one Marx articulated in his discussion of ideology, and it has been developed by feminist standpoint theorists. (For an overview and critical discussion of this work, see Harding 1986, 1991.) To see the world from the perspective of privilege is to see it through distorting lenses that produce views that naturalize or otherwise justify the privilege in question. Those who are privileged are not doomed to view the world only through those lenses, but it is at the very least unlikely that learning to see differently will come about without the experience of listening to and learning from those who are differently placed—an experience not likely without conscious effort. Preempting or stigmatizing those other voices is a central strategy of the ideology of privilege: either we think we already know what they have to say, or we think we have reasons for rejecting in advance the possibility that what they say will be of value.

María Lugones (in a talk at the University of Minnesota) gave a very interesting example of this sort of critique by looking at the moral virtue of integrity defined as singularity of subjectivity and of will, including a direct connection between one's will and one's actions. (For a related discussion, see Lugones 1990.) From the perspective of the subordinated, there are two things that are evident. One is the visibility of the meant-to-be invisible facilitation provided by various kinds of social structures that step in and enable the intentions of the privileged to issue in actions that appear to have been done in a straightforward, direct, unmediated fashion. Marilyn Frye (1983) gives an illuminating example of this phenomenon when she suggests we see patriarchy through the analogy with a play that proceeds thanks to the invisible labor of stagehands (women) who are not meant to be seen. The play of patriarchy is disrupted—revealed as a play—when women stop looking at the play and at the actors and start looking at each other.

The second thing that, on Lugones's account, becomes evident from the perspective of the subordinated is the necessity on their part of various sorts of ad hoc devices compensating for the meant-to-be invisible facilitations provided to the privileged by the social structure. Consider, for example, the ways in which language typically allows the privileged to say directly and straightforwardly what they mean. Part of the reason that it is so easy to ridicule attempts at bias-free language reform is that it really is awkward and graceless to avoid, for example, using the masculine as though it were generic. What we should learn from this discovery is just how pervasive and pervasively debilitating the biases in our languages are, how much they are set up to make some able to navigate smoothly while others systematically founder on jagged assaults to their self-esteem or engage in constant tacking against the wind in order to avoid them.

Similar situations arise in the realm of action. The privileged can (apparently) directly and straightforwardly do what they intend ("apparently" because we are not supposed to notice the assistance they are receiving, the ways in which all the paved roads are laid out to go where they want to go). The subordinated are driven to constructing various Rube Goldberg devices—in the realm of discourse, circumlocution; in the realm of action, manipulation. That is, everything that they do to make things happen, all the bits and pieces, are both patched together in a makeshift fashion and are also glaringly evident. They are "devious," not "straightforward," or so it seems from the perspective of privilege.

The articulation of such critical perceptions and their entry into the moral conversations of the privileged require what Lugones (1987) refers to as "world travel," and various things go into effecting it. One is the cultivation of relationships of trust; another is the willingness of the privileged to acknowledge others' views, including others' views of *them*; another is the evocation and the articulation of outlaw emotions and their acknowledgment as an epistemic resource; and a final one is attention to how we make each other up, especially across lines of privilege, how we create the possibilities of meaningfulness in each other's lives.

A crucial piece of such epistemic work is to transform dismissible cynicism and rage on the part of the subordinated into critique and transformation. As long as people feel effectively shut out of the procedures of moral conversation, they are likely to respond, if they respond at all, with global cynicism and inchoate rage, which are too easily dismissible, rather than with focused, specific, and intelligible critique, which has the possibility of being transformational. The work of creating a context in which this articulation can occur is in part the task of the more privileged, as it is the structures of privilege that have set the problem, and it cannot be solely up to those who are subordinated to take on the remedial education of those who continue to benefit from their subordination.

The progress toward moral objectivity is thus in the form of a spiral: as we become better at the forms of emotional engagement that allow the emotional resources, in particular of those whose perspectives have been subordinated, to become articulated and to be taken seriously by themselves as well as by the relatively privileged, then the terms of the descriptive apparatus for our emotional lives shift, which then allows for even more articulation of emotionally informed perceptual responses, and so on. What appeared to be a problematic circle is in fact a spiral, and if we have defined objective judgments as those that we have good reason to believe have been subjected to effective critique, existing in contexts that allow for the future possibility of further critique—that is, judgments that we have good reason to believe are stable across a wide range of different perspectives and that will shift, if they do, not capriciously but intelligibly—then it follows that the social constructedness of emotions and the ineliminable role of emotions in moral judgment work together in order to provide the possibility of moral objectivity.

Notes

1. For an extensive and very thoughtful discussion based on this (mis)reading of the argument, see Grimshaw (1986).
2. There is an extensive literature on what constitutes a disease, much of it concerned with arguing that what is called the "medical model" misidentifies as disease complexes of behavior and symptoms that take their significance from their social context. See, for example, Zita (1988).
3. This point is similar to the one Helen Longino makes in arguing for the importance of attending to "the specification, or constitution, of the object of inquiry" (Longino 1990, 99); she in turn acknowledges Michel Foucault and Donna Haraway.

4. I argue this claim, concerning all complex mental states, more fully in several of the essays in Scheman (1993) and in an unpublished manuscript, "Types, Tokens, and Conjuring Tricks."
5. Thanks to Daniel Hurwitz for deepening my reading of the sonnet, in particular, in insisting that the point cannot be that "it" simply is not love.
6. I discuss this example more fully in Scheman (forthcoming).
7. A much more subtle objection to the role that I give others in the identification of feelings (and a persuasive argument for serious attention to the inchoate) is explored in Sue Campbell's doctoral dissertation in philosophy at the University of Toronto, 1993, a revised version of which I understand will be forthcoming as a book.
8. I am indebted here to the account in Longino (1990) of objectivity as primarily a characteristic of scientific practice, rather than of theories or whatever else result from that practice, which practice is objective insofar as it embodies the conditions of effective critique, so as to take into account the essential role played by values. Presumably no one would argue that moral knowledge should be value free, but the problems of how to characterize objectivity in the face of diversity are not dissimilar.

References

Campbell, Susan. 1993. *Expression and the Individuation of Feeling*. Ph.D dissertation, University of Toronto.

Frye, Marilyn. 1983. "To Be and Be Seen: The Politics of Reality." In *The Politics of Reality*. Trumansburg, N.Y.: Crossing Press.

Grimshaw, Jean. 1986. *Philosophy and Feminist Theory*. Minneapolis: University of Minnesota Press.

Harding, Sandra. 1986. *The Science Question in Feminism*. Ithaca: Cornell University Press.

Harding, Sandra. 1991. *Whose Science? Whose Knowledge?* Ithaca: Cornell University Press.

Jaggar, Alison M. 1989. "Love and Knowledge: Emotion in Feminist Epistemology." In *Women, Knowledge, and Reality: Explorations in Feminist Epistemology*. Edited by Ann Garry and Marilyn Pearsall. Boston: Unwin Hyman.

Longino, Helen. 1990. *Science as Social Knowledge: Values and Objectivity in Scientific Inquiry*. Princeton: Princeton University Press.

Lovibond, Sabina. 1983. *Realism and Imagination in Ethics*. Minneapolis: University of Minnesota Press.

Lugones, María. 1987. "Playfulness, 'World'-Travel, and Loving Perception." *Hypatia* 2:3–19.

Lugones, María. 1990. "Hispaneando y Lesbiando: On Sarah Hoagland's *Lesbian Ethics*." *Hypatia* 5:138–146.

Lugones, María. 1991. "On the Logic of Pluralist Feminism." In *Feminist Ethics*. Edited by Claudia Card. Lawrence: University of Kansas Press.

Rich, Adrienne. 1979. "Women and Honor: Notes on Lying." In *On Lies, Secrets, and Silence*. New York: Norton.

Scheman, Naomi. 1980. "Anger and the Politics of Naming." In *Women and Language in Literature and Society*. Edited by Sally McConnell-Ginet, Ruth Borker, and Nellie Furman. Westport, Conn.: Praeger.

Scheman, Naomi. 1983. "Individualism and the Objects of Psychology." In *Discovering Reality: Feminist Perspectives on Epistemology, Metaphysics, Methodology, and the Philosophy of Science*. Edited by Sandra Harding and Merrill B. Hintikka. Dordrecht: Reidel.

Scheman, Naomi. 1993. *Engenderings: Constructions of Knowledge, Authority, and Privilege.* New York: Routledge.

Scheman, Naomi. Forthcoming. "Forms of Life: Mapping the Rough Ground." In *The Cambridge Companion to Wittgenstein.* Edited by Hans Sluga and David Stern. Cambridge: Cambridge University Press.

Wittgenstein, Ludwig. 1953. *Philosophical Investigations.* New York: Macmillan.

Zita, Jacquelyn. 1988. "The Premenstrual Syndrome: 'Dis-easing' the Female Cycle." *Hypatia* 3:77–99.

Part IV

Agency and Responsibility

Chapter 13

Justifying Morality and the Challenge of Cognitive Science

James P. Sterba

The task of trying to justify morality by showing that morality is required by rationality is not a task that many contemporary philosophers have set for themselves. No doubt most of these philosophers would like to have an argument showing that morality is rationally required and not just rationally permissible because if morality is just rationally permissible, then egoism and immorality would be rationally permissible as well. But given the history of past failures to provide a convincing argument that morality is rationally required, most contemporary philosophers have simply given up hope of defending morality in this way. In this chapter I propose to provide an argument that shows that morality is rationally required and then consider the challenges that recent developments in cognitive science raise to this form of moral argumentation.

Justifying Morality

Let us begin by imagining that we are members of a society deliberating over what sort of reasons for action we should accept. Let us assume that each of us is capable of entertaining and acting upon both self-interested and moral reasons and that the question we are seeking to answer is what sort of reasons for action it would be rational for us to accept.[1] This question is not about what sort of reasons we should publicly affirm since people will sometimes publicly affirm reasons that are quite different from those they are prepared to act upon; rather it is a question of what reasons it would be rational for us to accept at the deepest level—in our heart of hearts.

Of course, there are people who are incapable of acting upon moral reasons. For such people, there is no question about their being required to act morally or altruistically. Yet the interesting philosophical question is not about such people but about people, like ourselves, who are capable of acting self-interestedly or morally and are seeking a rational justification for following a particular course of action.

In trying to determine how we should act, we would like to be able to construct an argument for morality that does not beg the question against rational egoism. The question here is what reasons each of us should take as supreme, and this question would be begged against rational egoism if we propose to answer it simply by assuming from the start that moral reasons are the reasons that each of us should take as supreme. But the question would be begged against morality as well if we proposed to answer the question simply by assuming from the start that self-interested reasons are the reasons that each of us should take as supreme. This means, of course, that we cannot answer the question of what reasons we should take as supreme simply by assuming the general principle of egoism: Each person ought to do what best serves his or her overall self-interest. We can no more argue for egoism simply by denying the relevance of moral reasons to rational choice than we can argue for pure altruism simply by denying the relevance of self-interested reasons to rational choice and assuming the following principle of pure altruism: Each person ought to do what best serves the overall interest of others. Consequently, in order not to beg the question against either egoism or altruism, we have no other alternative but to grant the prima facie relevance of both self-interested and moral reasons to rational choice and then try to determine which reasons we would be rationally required to act upon, all things considered.

Here it might be objected that we do have non-question-begging grounds for favoring self-interested reasons over moral reasons, if not egoism over altruism. From observing ourselves and others, don't we find that self-interested reasons are better motivators than are moral reasons, as evidenced by the fact that there seem to be more egoistically inclined people in the world than there are altruistically inclined people? It might be argued that because of this difference in motivational capacity, self-interested and moral reasons should not both be regarded as prima facie relevant to rational choice.

But is there really this difference in motivational capacity? Do human beings really have a greater capacity for self-interested behavior than for moral or altruistic behavior? If we focus for a change on the behavior of women, I think we are likely to observe considerable more altruism than egoism among women, particularly with respect to the care of their families.[2] Of course, if we look to men, given the prevailing patriarchal social structures, we may tend to find more egoism than altruism.[3] But surely any differences that exist between men and women in this regard are primarily due to the dominant patterns of socialization—nurture rather than nature.[4] In any case, it is beyond dispute that we humans are capable of both self-interested and altruistic behavior, and given that we have these capabilities, it seems reasonable to ask which ones should have priority.

Our situation is that we find ourselves with some capacity to move along a spectrum from egoism to pure altruism, with someone like Mother Teresa of Calcutta representing the paradigm of pure altruism and someone like Thrasymachus of Plato's *Republic* representing the paradigm of egoism. Obviously, our ability to move along this spectrum will depend on our starting point, the strength of our habits, and the social circumstances under which we happen to be living. But at the outset, it is reasonable to abstract from these individual variations and simply focus on the general capacity virtually all of us have to act on both self-interested and moral reasons. From this, we should conclude that both sorts of reasons are relevant to rational choice and then ask the question which reasons should have priority. Later, with this question answered, we can take into account individual differences and the effects of socialization to adjust our expectations and requirements for particular individuals and groups. Initially, however, all we need to recognize is the relevance of both self-interested and altruistic reasons to rational choice. In this regard, there are two kinds of cases that must be considered: cases in which there is a conflict between the relevant self-interested and moral reasons and cases in which there is no such conflict.[5]

It seems obvious that where there is no conflict, and both reasons are conclusive reasons of their kind, both reasons should be acted upon. In such contexts, we should do what is favored by both morality and self-interest.

Consider the following example. Suppose you accepted a job marketing a baby formula in underdeveloped countries, where the formula was improperly used, leading to increased infant mortality.[6] Imagine that you could just as well have accepted an equally attractive and rewarding job marketing a similar formula in developed countries, where the misuse does not occur, so that a rational weighing of the relevant self-interested reasons alone would not have favored your acceptance of one of these jobs over the other.[7] At the same time, there were obviously moral reasons that condemned your acceptance of the first job—reasons that you presumably are or were able to acquire. Moreover, by assumption in this case, the moral reasons do not clash with the relevant self-interested reasons; they simply made a recommendation where the relevant self-interested reasons are silent. Consequently, a rational weighing of all the relevant reasons in this case could not but favor acting in accord with the relevant moral reasons.[8]

Needless to say, defenders of rational egoism cannot but be disconcerted with this result since it shows that actions that accord with rational egoism are contrary to reason at least when there are two equally good ways of pursuing one's self-interest, only one of which does not conflict with the basic requirements of morality. Notice also that in cases where there are two equally good ways of fullling the basic requirements of morality, only one of which does not conflict with what is in a person's overall self-interest, it is not at all disconcerting for defenders of morality to admit that we are rationally required to choose the way that does not conflict with what is in our overall self-interest. Nevertheless, exposing this defect in rational egoism for cases where moral reasons and self-interested reasons do not conflict would be but a small victory for defenders of morality if it were not also possible to show that in cases where such reasons do conflict, moral reasons would have priority over self-interested reasons.

When we rationally assess the relevant reasons in such conflict cases, it is best to cast the conflict not as one between self-interested reasons and moral reasons but instead as one between self-interested reasons and altruistic reasons.[9] Viewed in this way, three solutions are possible. First, we could say that self-interested reasons always have priority over conflicting altruistic reasons. Second, we could say just the opposite, that altruistic reasons always have priority over conflicting self-interested reasons. Third, we could say that some kind of a compromise is rationally required. In this compromise, sometimes self-interested reasons would have priority over altruistic reasons, and sometimes altruistic reasons would have priority over self-interested reasons.

Once the conflict is described in this manner, the third solution can be seen to be the one that is rationally required. This is because the first and second solutions give exclusive priority to one class of relevant reasons over the other, and only a completely question-begging justification can be given for such an exclusive priority. Only the third solution, by sometimes giving priority to self-interested reasons and sometimes giving priority to altruistic reasons, can avoid a completely question-begging resolution.

Consider the following example. Suppose you are in the waste disposal business and have decided to dispose of toxic wastes in a manner that is cost-efficient for you but predictably causes significant harm to future generations. Imagine that there are alternative methods available for disposing of the waste that are only slightly less cost-efficient and will not cause any significant harm to future generations.[10] In this case, you are to weigh your self-interested reasons favoring the most cost-efficient disposal of the toxic wastes against the relevant altruistic reasons favoring the avoidance of significant harm to future generations. If we suppose that the projected loss of benefit to yourself was ever so slight and the projected harm to future generations was ever so great, then a nonarbitrary compromise between the relevant self-interested and altruistic reasons would have to favor the altruistic reasons in this case. Hence, as judged by a non-question-begging standard of rationality, your method of waste disposal is contrary to the relevant reasons.

Notice also that this standard of rationality will not support just any compromise between the relevant self-interested and altruistic reasons. The compromise must be a nonarbitrary one, for otherwise it would beg the question with respect to the opposing egoistic and altruistic views. Such a compromise would have to respect the rankings of self-interested and altruistic reasons imposed by the egoistic and altruistic views, respectively. Since for each individual there is a separate ranking of that individual's relevant self-interested and altruistic reasons, we can represent these rankings from the most important reasons to the least important reasons, as follows:

Individual A		Individual B	
Self-interested reasons	Altruistic reasons	Self-interested reasons	Altruistic reasons
1	1	1	1
2	2	2	2
3	3	3	3
.	.	.	.
.	.	.	.
.	.	.	.

Accordingly, any nonarbitrary compromise among such reasons in seeking not to beg the question against egoism or altruism will have to give priority to those reasons that rank highest in each category. Failure to give priority to the highest-ranking altruistic or self-interested reasons would, other things being equal, be contrary to reason.

Of course, there will be cases in which the only way to avoid being required to do what is contrary to your highest-ranking reasons is by requiring someone else to do what is contrary to her highest-ranking reasons. Some of these cases will be "lifeboat cases." But although such cases are surely difficult to resolve (maybe only a chance mechanism can offer a reasonable resolution), they surely do not reflect the typical conflict between the relevant self-interested and altruistic reasons that we are or were able to acquire, for typically one or the other of the conflicting reasons will rank higher on its respective scale, thus permitting a clear resolution.

It is important to see how morality can be viewed as just such a nonarbitrary compromise between self-interested and altruistic reasons. First, a certain amount of self-regard is morally required or at least morally acceptable. Where this is the case, high-ranking self-interested reasons have priority over the low-ranking altruistic reasons. Second, morality obviously places limits on the extent to which people should pursue their own self-interest. Where this is the case, high-ranking altruistic reasons have priority over low-ranking self-interested reasons. In this way, morality can be seen to be a nonarbitrary compromise between self-interested and altruistic reasons, and the "moral reasons" that constitute that compromise can be seen as having an absolute priority over the self-interested or altruistic reasons that conflict with them.

Now it might be objected that although the egoistic and the altruistic views are admittedly question begging, the compromise view is equally so and, hence, in no way preferable to the other views. In response, I deny that the compromise view is equally question begging when compared with the egoistic and altruistic views but then concede that the view is to a lesser degree question begging nonetheless. For a completely non-question-begging view starts with assumptions that are acceptable to all sides of a dispute. However, the assumption of the compromise view that the highest-ranking altruistic reasons have priority over conflicting lower-ranking self-interested reasons is not acceptable from an egoistic perspective. Nor is the compromise view's assumption that the highest-ranking self-interested reasons have priority over conflicting lower-ranking altruistic reasons acceptable from an altruistic perspective. So one part of what the compromise view assumes about the priority of reasons is not acceptable from an egoistic perspective, and another part is not acceptable from an altruistic perspective, and, hence, to that extent the compromise view does beg the question against each view. Nevertheless, since the whole of what egoism assumes about the priority of reasons is unacceptable from an altruistic perspective and the whole of what altruism assumes about the priority of reasons is unacceptable from an egoistic perspective, each of these views begs the question against the other to a far greater extent than the compromise view does against each of them. Consequently, on the grounds of being the least question begging, the compromise view is the only nonarbitrary resolution of the conflict between egoism and altruism.

Notice, too, that this defense of morality succeeds not only against the view that rational egoism is rationally preferable to morality but also against the view that rational egoism is only rationally on a par with morality. The "weaker view" does not claim that we all ought to be egoists. Rather it claims that there is just as good reason for us to be egoists as to be pure altruists or anything in between. As Kai Nielson summarizes this view, "We have not been able to show that reason requires the moral point of view or that all really rational persons not be individual egoists. Reason doesn't decide here."[11] Yet since the above defense of morality

shows morality to be the only nonarbitrary resolution of the conflict between self-interested and altruistic reasons, it is not the case that there
is just as good reasons for us to endorse morality as to endorse rational
egoism or pure altruism. Thus, the above defense of morality succeeds
against the weaker as well as the stronger interpretation of rational
egoism.

It might be objected that this defense of morality could be undercut
if in the debate over egoism, altruism, and morality we simply give up
any attempt to provide a non-question-begging defense for any one of
these views. But we cannot rationally do this, because we are engaged
in this debate as people who can act self-interestedly, can act altruistically,
can act morally, and we are trying to discover which of these ways
of acting is rationally justied. To rationally resolve this question, we have
to be committed to deliberating in a non-question-begging manner as far
as possible. So as far as I can tell, there is no escaping the conclusion that
morality can be given a non-question-begging defense over egoism and
altruism.

This approach to defending morality has been generally neglected by
previous moral theorists. The reason is that such theorists have tended to
cast the basic conflict with rational egoism as one between morality and
self-interest. For example, according to Kurt Baier, "The very *raison d'etre*
of a morality is to yield reasons which overrule the reasons of self-interest
in those cases when everyone's following self-interest would be harmful to
everyone." [12] Viewed in this light, it did not seem possible for the defender
of morality to be supporting a compromise view, for how could such a
defender say that, when morality and self-interest conflict, morality should
sometimes be sacriced for the sake of self-interest? But while previous
theorists understood correctly that moral reasons could not be compromised in favor of self-interested reasons, they failed to recognize that this
is because moral reasons are already the result of a nonarbitrary compromise between self-interested and altruistic reasons. Thus, unable to
see how morality can be represented as a compromise solution, previous
theorists have generally failed to recognize this approach to defending
morality.

This failure to recognize that morality can be represented as a compromise between self-interested and altruistic reasons also helps explain
Thomas Nagel's inability to find a solution to the problem of the design of
just institutions. [13] According to Nagel, to solve the problem of the design
of just institutions, we need a morally acceptable resolution of the conflict
between the personal and the impersonal standpoints, which Nagel thinks
is unattainable. But while Nagel may be right that a morally acceptable
resolution of the conflict between these two standpoints is unattainable,

the reason why this may be the case is that these two standpoints already represent different resolutions of the conflict between self and others. The personal standpoint represents the personally chosen resolution of this conflict, while the impersonal standpoint represents a completely impartial resolution of this conflict, which may not be identical with the personally chosen resolution. Since each of these standpoints already represents a resolution of the conflict between oneself and others, any further resolution of the conflict between the two standpoints would seem to violate the earlier resolutions either by favoring oneself or others too much or not enough. It is no wonder, then, that an acceptable resolution of the two standpoints seems unattainable. By contrast, if we recast the underlying conflict between oneself and others, as I have suggested, in terms of a conflict between egoism and altruism, self-interested reasons and altruistic reasons, then happily a non-question-begging resolution can be seen to emerge.

The Challenge of Cognitive Science

I now turn to consider some of the challenges of cognitive science to this form of moral argumentation. In his recent book, *Moral Imagination*, Mark Johnson reports:

> Psychologists, linguists and anthropologists have discovered that most categories used by people are not actually definable by a list of features. Instead, people tend to define categories (e.g., bird) by identifying certain prototypical members of the category (e.g., robin), and they recognize other nonprototypical members (e.g., chicken, ostrich, penguin) that differ in various ways from the prototypical ones. There is seldom any set of necessary and sufficient features possessed by all the members of the category. In this way our ordinary concepts are not uniformly or homogeneously structured.[14]

According to Johnson and Stephen Stich, this empirical research conflicts with certain standard assumptions of traditional and contemporary moral theory.[15]

Stich elicits some of these assumptions from an analysis of two examples.[16] The first is from Plato's *Euthythro:*

Socrates.   I abjure you to tell me the nature of piety and impiety, which you say that you know so well, and of murder, and of other offenses against the gods. What are they? Is not piety in every action always the same?

Euthyphro.   To be sure, Socrates.

Socrates.   And what is piety, and what is impiety?... Tell me what is the nature of this idea, and then I shall have a standard to which I may look, and by which I may measure actions, whether yours or those of any one else, and then I shall be able to say that such and such an action is pious, such another impious.

Euthyphro.   I will tell you, if you like.... Piety... is that which is dear to the gods, and impiety is that which is not dear to them.

Socrates.   Very good, Euthyphro; you have now given me the sort of answer which I wanted. But whether what you say is true or not I cannot as yet tell....

The quarrels of the gods, noble Euthyphro, when they occur, are of a like nature [to the quarrels of men].... They have differences of opinion... about good and evil, just and unjust, honorable and dishonorable....

Euthyphro.   You are quite right.

Socrates.   Then, my friend, I remark with surprise that you have not answered the question which I asked. For I certainly did not ask you to tell me what action is both pious and impious: but now it would seem that what is loved by the gods is also hated by them.

According to Stich, a standard assumption of traditional moral theory that can be elicited from this passage is that a correct definition must provide individually necessary and jointly sufficient conditions for the application of the concept being defined.

In his second example, Stich reports on his efforts to get his students to make explicit the reasons why they think "that there is nothing wrong with raising cows, pigs, and other common farm animals for food." Stich asked his students to suppose that a group of very wealthy gourmets decide that it would be pleasant to dine occasionally on human flesh. To achieve this goal, they hire a number of couples who are prepared to bear infants to be harvested for the table. When his students objected because human infants have the potential to become full-fledged human adults, Stich altered the example so that the women were impregnated with sperm that resulted in only severely retarded infants. And when his students still objected because the infants were human, Stich asks them to consider a group of Icelanders who are so genetically different from us as to render them a different species, and asks whether there is any objection to eating them. Stich maintains that his classroom endeavor of seeking morally relevant differences between harvesting human babies and harvesting non-humans is similar to the Platonic endeavor to find definitions because both endeavors seek to specify the use of categories by necessary and sufficient conditions.

By contrast, Stich claims that a relevant lesson we should learn from research in cognitive science is that "a concept consists of a set of salient features or properties that characterize only the best of 'prototypical' members of a category. This prototype representation will, of course, contain a variety of properties that are lacking in some members of the category ... On the prototype view of concepts, objects are classified as members of a category if they are sufficiently similar to the prototype —that is, if they have sufficient number of properties specified in the prototype representation."[17] Yet how is this understanding of how we use concepts supposed to provide a better account of what is going on in Socrates' discussion with Euthyphro and in Stich's discussion with his students? In the case of Socrates' discussion with Euthyphro, are we to regard things that are loved by the gods as prototypical examples of piety? But what if, as in Socrates' discussion with Euthyphro, we are dealing with things that are loved by some gods and hated by others? Are these things prototypical examples of piety? Presumably not.

What, then, is a prototypical example of piety? Maybe, in Socrates' discussion, it is something that is loved by some of the gods and not hated by any of them. And maybe a prototypical property of piety is the property of being loved by some of the gods and not hated by any of them. This might an acceptable interpretation of Socrates' discussion, but why couldn't we interpret Plato as making the same point without employing the terminology of prototypical examples and prototypical properties? Why couldn't we interpret Socrates' discussion with Euthyphro as a search for a better definition or standard of piety without assuming that that definition or standard must be expressible in necessary and sucient conditions? Surely Socrates' attempt in *The Republic* to capture what is a just person by analogy with what is a just state does not look like an attempt to define justice through necessary and sufficient conditions.

Now let us turn to the example of Stich's discussion with his students. In this example, what are prototypical examples of what we can eat and cannot eat? Is it clear that a pig is a prototypical example of what we can eat and a normal human baby a prototypical example of what we cannot eat? According to Stich, a prototypical example has salient features or properties that characterize only the best of prototypical members of a category. But what are these features? In the case of normal adult humans, it is the features of being a human and having actually and potentially a high level of intelligence and self-awareness. In the case of a normal human baby, presumably it is the features of being a human and having the potential for a high level of intelligence and self-awareness. In the case of severely defective human babies, it is the feature of simply being human, and, in the case of Stich's hypothetical Icelanders, it is the feature of actually having a high level of intelligence and self-awareness without being human. Pigs and the other nonhuman animals that we eat, have none of these prototypical features.

Granted that this is roughly how we actually employ our concept of things we cannot eat, how does knowing this help in a discussion of animal rights? Obviously, knowing how we use the concept of things we cannot eat does not, by itself, tell us how we should use this concept if we want to have a morally defensible view. To determine how we should use the concept in order to have a morally defensible view, we have to consider what should be prototypical examples of things that we cannot eat. At this juncture, it becomes an open question whether we should include pigs and other nonhuman animals in that category, either because of their actual level of intelligence or because we have available other nonsentient resources for food. This seems to me what Stich's discussion of animals rights was all about in the first place—not how we use our concepts but how we should use them.

Of course, Stich might accept this characterization of his discussion but then claim that any attempt to discover how we should use such concepts is ordinarily conceived as an attempt to discover necessary and sufficient conditions. But why should this pursuit be so construed? Suppose we were to modify our concept of things we cannot eat to include any intelligent nonhuman or even to include any living thing that we did not need to eat to have a decent life. Wouldn't defenders of animal rights be content with this modification of our concept, even though it is not expressible in terms of necessary and sufficient conditions? I think that what defenders of animal rights are interested in doing is simply enlarging the class of things we cannot eat. Accordingly, they are not concerned to reject any morally defensible prototypical structure of the items to be included within that class.

In his recent book, Mark Johnson also pursues an analogous argument against traditional and contemporary moral theory—what he calls "moral law folk theory." By moral law folk theory, Johnson means any view that regards moral reasoning as consisting entirely of bringing concrete cases under moral laws or rules that specify the "right thing to do" in a given instance.[18] Johnson further claims that according to moral law folk theory, applying moral laws or rules to concrete situations "is simply a matter of determining whether the necessary and sufficient conditions that capture the concepts contained in these laws actually obtain in practice."[19] Johnson regards Immanuel Kant's ethical theory as a traditional articulation of moral law folk theory and Alan Donagan's ethical theory as a contemporary articulation of the theory.

It is not clear, however, that Kant's ethical theory conforms to Johnson's model of moral law folk theory. For one thing, Kant seems interested in doing much more than moral law folk theory allows. For example, Kant is concerned to articulate the presumptions of practical (moral) reason and show how these presumptions are possible. But as Johnson characterizes moral law folk theory, this is no part of its agenda. For another thing, Kant does not think that the application of moral laws and rules to everyday life is an easy manner, even for those who are conscious of such laws and rules.

Johnson quotes the following passage from Kant's *Critique of Practical Reason:* "Since all instances of possible actions are only empirical and can belong only to experience and nature, it seems absurd to wish to find a case in the world of sense, and thus standing under a law of nature, which admits the application of a law of freedom to it, and to which we could apply the supersensuous moral ideal of the morally good, so that the latter could be exhibited in concreto."[20] Yet Johnson does not conclude from this passage that for Kant the application of moral laws and rules is not a straightforward deduction of particular moral requirements from universal laws, and, hence, represents an account of moral reasoning that is similar to the account that Johnson wants to defend. Rather Johnson concludes from this passage that "unless Kant can show how a pure formal principle can apply to experience, his entire project of purely rational morality is undermined."[21] This is like saying that Kant's view must involve a straight-forward deduction of particular requirements from universal laws, appearances to the contrary.

Actually, concluding from this passage that for Kant the application of general laws and rules to practice can be quite difficult makes more sense, and it also makes more sense of what Kant says elsewhere: that the faculty of judgment can be improved through the use of examples—examples of roles' being applied in concrete cases. According to Kant, "Examples are thus the go-cart of judgement: and those who are lacking in the general talent [of applying rules] can never dispense with them."[22]

Possibly Johnson would allow that his characterization of Kant as a paradigm moral law folk theorist is somewhat strained. This may explain why Johnson moves on to a discussion of Alan Donagan's theory, which Johnson considers to be "the most representative, well argued and sophis-ticated version of rule (Moral Law) theory."[23]

The basic principle of Donagan's theory is that "it is impermissible not to respect every human being, oneself and any other, as a rational crea-ture." Johnson's criticism of Donagan's view centers on how this principle is to be interpreted and applied. First, Johnson criticizes Donagan's claim that the principle can be understood in itself. Rather, according to Johnson, the principle "has its meaning only in relation to a culture's shared evolv-ing experience and social interactions."[24] Second, Johnson claims that the concept of a person, which is at the heart of Donagan's basic moral princi-ple, is a complex concept with a prototypical structure. According to Johnson, this concept has as prototypical instances (for example, sane, adult, white, heterosexual males) surrounded by nonprotypical instances (for example, females, nonwhites, children, senile elderly, and mentally disabled). Johnson also considers the possibility of extending the concept of a person to include nonhuman animals and even the ecosystem of the planet.[25] Let us consider each of these criticisms of Donagan's theory in turn.

Johnson's first criticism of Donagan fails to hold because, as the passage from Donagan's work that Johnson quotes just before raising this criticism makes clear, Donagan actually holds virtually the same view about the application of moral rules that Johnson himself adopts. Thus, on page 90, Johnson says that central moral concepts cannot be understood in themselves but "only in relation to a culture's shared, evolving experience and social interactions." But on page 89, Johnson quotes Donagan as saying, "The fundamental concept of respecting every human being as a rational creature is fuzzy at the edges in the superficial sense that its application to this or that species of case can be disputed. But among those who share in the life of a culture in which the Hebrew-Christian moral tradition is accepted, the concept is in large measure understood in itself." Since both Johnson and Donagan affirm that central moral concepts are understood within a cultural setting, I do not see any difference between Johnson and Donagan's views here.

With regard to Johnson's second criticism of Donagan's theory, once we recognize that for both authors our understanding of central moral concepts derives from the cultural tradition in which we live, there is no reason why Donagan need reject Johnson's idea that our concept of a person has a prototypical structure. In fact, in his book *The Theory of Morality*, we find Donagan maintaining that "in the first instance, respect is recognized as owed to beings by virtue of a state they are in: say, that of rational agency. If there are beings who reach that state by a process of development natural to normal members of their species, given normal nurture, must not respect logically be accorded to them, whether they have reached that state or not?"[26] So we find Donagan endorsing here something like the same prototypical structure for the concept of a person that Johnson advocates.

There is, of course, the need to distinguish, as I did earlier, between how we actually use the concept of a person and how we should use that concept. Here, we can agree with Johnson that as we actually use the concept, it has just the prototypical structure that he describes with sane, adult, white, heterosexual males as prototypical instances of persons surrounded by nonprototypical instances of females, homosexuals, nonwhites, children, senile elderly, and mentally disabled.[27] But there remains the question of how we should use this concept, and here there seem to be good reasons for expanding the class of prototypical instances of persons or, more broadly, things that are ends in themselves to include not only all human beings but possibly all living beings as well. With all this, Johnson may well agree, but as far as I can tell, there is nothing in Donagan's view to keep him from agreeing as well.

More recently, Paul M. Churchland has put the challenge of cognitive science to traditional and contemporary moral philosophy in a particularly forceful manner.[28] According to Churchland, developments in cognitive science suggest that a person's moral capacity has less to do with the application of rules than with the recognition of the appropriateness of certain learned prototypes. He illustrates the point with the example of abortion during the first trimester. Churchland claims that one side in the abortion debate in considering the status of the early fetus invokes the moral prototype of an innocent person, while the other side addresses the same situation and invokes the prototype of an unwanted bodily growth. So the moral argument over abortion, according to Churchland, should be viewed as a debate concerning the appropriateness of one or the other of these competing prototypes.

Yet exactly how is this way of viewing the problem of abortion different from trying to determine whether the rule "Protect innocent persons" or the rule "Do whatever you desire with unwanted bodily growths" justifiably applies in this context? Why should we have to choose between thinking about moral problems in terms of either recognizing the appropriate prototypes or applying the appropriate rules? Why couldn't it be both?

Churchland resists this conclusion because he wishes to view the exercise of our moral capacity as akin to normal perception. Failure to exercise our moral capacity, according to Churchland, is a failure to see things as a normal person would see them. He writes,

> No one is perfect. But some people, as we know, are notably less perfect than the norm, and their failures are systematic. In fact, some people are rightly judged to be thoughtless jerks, chronic troublemakers, terminal narcissists, worthless scumbags and treacherous snakes; not to mention bullies, thieves and vicious sadists. Whence stem these sorry failings?
>
> From many sources, no doubt. But we may note right at the beginning that a simple failure to develop the normal range of moral *perception* and social *skills* will account for a great deal here (author's italics).[29]

The obvious problem with tying morality to how most people think about the matter, as Churchland does here, is that most people can be wrong about certain moral issues; in fact, most people can be wrong about many moral issues. Churchland's approach to morality equates morality with conventional morality, what is morally justied with what (most) people think is morally justied. No doubt cognitive science will prove a very useful tool for increasing our knowledge of conventional morality. This may be what Churchland has in mind when he suggests that moralists who agree with him will spend their time "mostly in the empirical trenches."[30] But getting clear about what most people think is moral is only part of the task of determining what is moral. In addition, we need to evaluate the reasons why people think that something is moral, whether they are well informed about the issue, and to what degree they have seriously reflected on the issue by examining alternative perspectives. All of this can be quite difficult to do. While noting what is the conventional morality on particular issues may not be a bad place to start if one is interested in what is truly moral, Churchland seems to think that it is the place to end as well. But equating morality with conventional morality could make sense only in a perfectly moral society, and surely this is not the society in which we presently live, and maybe it is not even a society in which we could reasonably expect to live in the future, assuming that the requirements of morality are fairly demanding, as I think they are. In that case, there will always be a large number of people, and, on some issues, a majority, who for reasons of self-interest have allowed themselves to be deceived about what they think are the requirements of morality.

This brings us to the task of assessing how my own justification for morality as presented in the first section of this chapter measures up to the challenges of cognitive science that I have raised in the second section. In particular, we need to determine whether my own justification for morality proceeds on the assumption that its key concepts can be defined in terms of necessary and sufficient conditions or that we can move deductively from morality so justified to specific moral requirements. As far as I can tell, the answer to both questions is no.

First, with respect to the characterization of morality that underlies my argument, no full definition of morality is presupposed. When one sets out to justify morality, one presupposes a context in which a justification for morality is appropriate or needed. In such contexts, the question of why be moral is raised, and a successful justification for morality responds to this question by providing oneself or others with adequate reasons for being moral. Since the reasons provided are usually reasons for being moral rather than being self-interested, the part of morality that is being justified is that which conflicts with self-interest. One way to put this is that we are focused on that part of morality, that set of altruistic reasons, which, within morality, have priority over conflicting self-interested reasons, and we are seeking to provide a non-question-begging justification for that priority. This is by no means a full characterization of morality, nor does it assume that a full characterization of morality can be given in terms of necessary and sufficient conditions.

Moreover, characterizing morality as a compromise between egoism and altruism does not make clear exactly how this compromise is to be carried out. There is no assumption that we can move deductively from morality so characterized to specific moral requirements. So we have libertarians with their moral ideal of liberty, welfare liberals with their moral ideal of fairness, communitarians with their moral ideal of the common good, feminists with their moral ideal of androgyny, and socialists with their moral ideal of equality all favoring different practical requirements. My own view is that when these different moral ideals are correctly interpreted, they all lead to the same practical requirements, which just happen to be those usually associated with a welfare liberal moral ideal: a right to welfare and a right to equal opportunity.[31] But even if I am right about this, there can be considerable debate as to how a right to welfare and a right to equal opportunity should be specified. In any case, there is no assumption that we can move deductively from morality characterized as a compromise between egoism and altruism to specific moral requirements.

What, then, are we to make of the challenge of cognitive science to both traditional and contemporary moral philosophy and my proposed justification for morality? I conclude by making an observation inspired by both feminism and peace studies. Too often doing philosophy is modeled after fighting a battle or making war. Arguments are attacked, proposals are killed, and positions are defeated. Given this model of doing philosophy, it may seem appropriate that as the findings of cognitive psychology are related to philosophy, they too must be capable of defeating or overturning certain cherished philosophical views. So in accord with this model of doing philosophy, advocates of cognitive science look for ways to defeat traditional and contemporary philosophical views. But why should warmaking be our model of doing philosophy? Why can't we have a more peaceful and cooperative model instead? Why can't we try to put the most favorable interpretation on the work of others rather than look for some interpretation with which we can disagree? Why can't we find points of agreement and attempt to build on the work of others rather than try to destroy their work and build anew? This is the approach I am recommending for relating the findings of cognitive science to traditional and contemporary moral philosophy. It has much to recommend it. At the very least, if you want others to agree with you, it cannot but help, where possible, to view their work in such a way that, to a large extent, they already do agree with you and you with them.

Notes

1. "Ought" presupposes "can" here. Unless the members of the society have the capacity to entertain and follow both self-interested and moral reasons for acting, it does not make any sense asking whether they ought or ought not to do so.

2. Nell Nodding, *Caring: A Feminine Approach to Ethics and Moral Education* (Berkeley: University of California Press, 1984); Joyce Tribilcot, ed., *Mothering* (Totowa, N.J.: Rowman and Littlefield, 1983); Susan Brownmiller, *Femininity* (New York: Linden Press, 1984).

3. James Doyle, *The Male Experience* (Dubuque, 1983); Marie Richmond-Abbot, ed., *Masculine and Feminine*, 2d ed. (New York: McGraw-Hill, 1991).

4. Victor Seidler, *Rediscovering Masculinity* (New York: Routledge, 1989); Larry May and Robert Strikwerda, *Rethinking Masculinity* (Lanham, Md.: Rowman and Littlefield, 1992).

5. Not all the reasons that people are or were able to acquire are relevant to an assessment of the reasonableness of their conduct. First, reasons that are evokable only from some logically possible set of opportunities are simply not relevant; the reasons must be evokable from the opportunities people actually possessed. Second, reasons that radically different people could have acquired are also not relevant. Instead, relevant reasons are those that people could have acquired without radical changes in their developing identities. Third, some reasons are not important enough to be relevant to a reasonable assessment of conduct. For example, a reason that I am able to acquire, which would lead me to promote my own interests or that of a friend just slightly more than I am now doing, is hardly relevant to an assessment of the reasonableness of my conduct. Surely I could not be judged as unreasonable for failing to acquire such a reason. Rather, relevant reasons are those that would lead one to avoid a significant harm to oneself (or others) or to secure a significant benefit to oneself (or others) at an acceptable cost to oneself (or others).

It is also worth noting that a given individual may not actually reflect on all the reasons that are relevant to deciding what she should do. In fact, one could do so only if one had already acquired all the relevant reasons. Nevertheless, reasonable conduct is ideally determined by a rational weighing of all the reasons that are relevant to deciding what one should do so that failing to accord with a rational weighing of all such reasons is to act contrary to reason.

6. For a discussion of the causal links involved here, see *Marketing and Promotion of Infant Formula in Developing Countries*, Hearing before the Subcommittee of International Economic Policy and Trade of the Committee on Foreign Affairs, U.S. House of Representatives, 1980. See also Maggie McComas et al., *The Dilemma of Third World Nutrition* (1983).

7. Assume that both jobs have the same beneficial effects on the interests of others.

8. I am assuming that acting contrary to reason is a significant failing with respect to the requirements of reason and that there are many ways of not acting in (perfect) accord with reason that do not constitute acting contrary to reason.

9. This is because, as I shall argue, morality itself already represents a compromise between egoism and altruism. So to ask that moral reasons be weighed against self-interested reasons is, in effect, to count self-interested reasons twice—once in the compromise between egoism and altruism and again when moral reasons are weighed against self-interested reasons. But to count self-interested reasons twice is clearly objectionable.

10. Assume that all these methods of waste disposal have roughly the same amount of beneficial effects on the interests of others.

11. Kai Nielson, "Why Should I be Moral? Revisited," *American Philosophical Quarterly* (1984): 90.

12. Kurt Baier, *The Moral Point of View*, abridged ed. (New York: Random House, 1965), p. 150.

13. Thomas Nagel, *Equality and Partiality* (Oxford: Oxford University Press, 1991).

14. Mark Johnson, *Moral Imagination* (Chicago: University of Chicago Press, 1993), pp. 8–9.

15. See Stephen Stich, "Moral Philosophy and Mental Representation," in *The Origin of Values*, ed. Michael Hechter (New York: Aldine De Gruyter, 1993), pp. 215–228.

16. Ibid., pp. 216–222.

17. Ibid., p. 223.

18. Johnson, *Moral Imagination*, p. 4.

19. Ibid., p. 81.

20. Ibid., p. 70.

21. Ibid., p. 70.

22. Immanuel Kant, *Critique of Pure Reason*, 1781 (= A), 134, and 1787 (= B) 174.

23. Johnson, *Moral Imagination*, p. 84.

24. Ibid., p. 90.

25. Ibid., pp. 97–98.

26. Alan Donagan, *The Theory of Morality* (Chicago: University of Chicago, 1977), p. 171.

27. Somehow homosexuals got omitted from Johnson's list of nonprototypical humans, but obviously they should not have been omitted.

28. In a paper presented by Paul Churchland at the Conference on Ethics and Cognitive Science, Washington University, St. Louis, April 8–10, 1994. The paper was exerpted from chapters 6 and 10 of Churchland's *The Engine of Reason, the Seat of the Soul: A Philosophical Essay on the Brain* (Cambridge, Mass.: MIT Press, 1995).

29. Ibid.

30. Ibid.

31. I have argued for this conclusion in *How to Make People Just* (Totowa, N.J.: Rowman and Littlefield, 1988).

Chapter 14

Moral Rationality

Susan Khin Zaw

What is moral rationality? Despite the fact that morality regulates practice, in practical philosophy currently the most popular answer aligns moral with speculative rationality on the nomological-deductive (N-D) model, that is, with *theoria*. On this account, moral rationality consists in submitting moral judgment to general moral principles derived from moral experience, for example, from moral intuitions about particular cases, much as scientific or theoretical reason subsumes particular phenomena under general laws. Practical moral reason discerns the right thing to do simply by reversing the process, deducing particular moral judgments from general principles. (For developed examples of this kind of account in the practical philosophical literature, see Glover 1982, chap. 2; Beauchamp and Childress 1989, chaps. 1, 2.)

Consistency is the measure of rationality so conceived; but notoriously, distinct but simultaneously tenable moral principles, unlike distinct but simultaneously tenable scientific laws, are expected often to yield inconsistent judgments on particular cases. However, such inconsistencies can be reduced by systematization of moral principles into a hierarchy of generality and abstraction, producing a moral decision procedure in cases of conflict between principles: conflicts at lower levels of the hierarchy can be settled by reference to principles of greater generality at higher levels. Systematization and a decision-procedure are goals also of some philosophical theories of morality. Moral rationality on this view is therefore held to increase with systematicity and hence with consistency in judgment, both of which can be increased by submitting judgment to the guidance of moral theory.

Systematic, decision-procedure-oriented moral theories have come under attack from within philosophy, but they have been attacked as accounts of morality, not as (implicit) accounts of moral rationality. Rather, critics of moral systematicity tend also to see moral rationalism—the view that reason is in some sense at the heart of morality—as an aspect or part of what is objectionable, for instance, as leading to a mistaken preoccupation with formal problems at the expense of moral content and motivation; it is these that should be the focus of philosophical attention, for example, in the form of philosophical accounts of the virtues. This objection to systematicity is thus consistent with acceptance of the N-D model as an account of rationality; the philosophical point at issue is how central it is to morality or how far it is applicable within it. This produces a reason-feeling, form-content, action-motivation polarization of philosophical opinion; that is, the deep questions about morality are conceived as forcing a philosophical choice at key points in the explanation between reason and feeling, form and content, action and motivation. But is it the nature of morality that forces us to such choices? Or are they forced on us by our conception of rationality?

The emergence of similar polarizations in other domains suggests the latter. Feminism provides an interesting example; there the reason-feeling polarization is manifest in the deep ideological divide between liberal rationalism and radical essentialism, generating political and theoretical conflict over the ground of femininity (social construction or biology?) and over what, if anything, should constitute feminist discourse. Should it confine itself to rational argument, or should it be a more discursive, more rhetorical and emotionally charged écriture feminine, as advocated by, for instance, Hélène Cixous (Cixous and Clément 1987)? Such disputes are the more bitter because the reason-feeling dichotomy aligns with the masculine-feminine dichotomy, so that feminists drawn toward the radical-essentialist pole perceive liberal rationalists as compromised by acceptance of masculine values (reason being perceived as masculine, feeling as feminine). Polar conflicts seem irresolvable because those drawn toward one pole distrust or despise what those drawn toward the other regard as indispensable.

A conception of rationality that gives rise to rationally insoluble problems leaves something to be desired, to say the least. If it gives rise to rationally insoluble problems fundamental to the domain, there is reason to believe it gravely flawed. This is perhaps the case within feminism; I shall argue that it is certainly the case within moral philosophy. In both, the fault line runs through the polarized problematic; for in both, this confronts the theoretically uncommitted with an unpalatable forced choice, or a series of related forced choices, between one pole and the other. The choice is unpalatable because most acknowledge some degree of attraction to both poles. Thus, the very existence of two poles of philosophical opinion, the persistence and deep-seatedness of whatever it is that makes both poles attractive, suggests that forcing a choice here is a mistake. Opinions polarize because both poles are such that there are strong philosophical grounds for clustering at them—hence, the apparent ineliminability of both philosophical disagreement over morality and political disagreement within feminism. In that case we cannot hope for progress toward consensus in either domain unless we can analyze out what it is that forces us to choose between the poles, what it is that generates that polarized problematic.

In what follows I shall argue that in morality, the main source of insoluble philosophical problems is too narrow a conception of reason. It is this that corrupts the problematic: we find ourselves faced with rationally insoluble problems because our conception of reason makes us ask the wrong questions. The way out of the impasse is to reject the problematic, deconstructing for that purpose the polarization that produces it. At this point feminist critiques are illuminating, for the polarizations of the moral and the feminist domains are deeply related. In both the way forward may be to reconceive reason in a way that makes a choice between the two poles unnecessary, which integrates reason and feeling, form and content, action and motivation, instead of separating them.

This project is less ambitious than it may appear, for there is a familiar form of reasoning whose job is precisely to bring these things together: practical reasoning (reasoning about what to do), of which moral reason is a species. The project is thus merely a reconsideration and repositioning of practical reason in relation to morality that allows a change of moral problematic, substituting soluble problems for at least some currently insoluble ones. Practical reason needs reconsideration because it is, I suggest, much less well understood than speculative or theoretical reason (reasoning about what is true), which is therefore generally seen as explanatorily prior, so that understanding of the processes of practical reason is sought by recourse to theoretical reason (e.g., via games theory, decision theory, probability theory, deontic logic). We know well enough what practical reason does—it devises means to ends—but often we do not know how it does it. Hence rational practice is identified with the importation of theory: rational medicine is scientific medicine, treatment directed by scientific understanding of disease. The example generalizes to a conception of rational practice as theoretically based technology. The N-D model of moral rationality is a case in point.

I shall argue that this model of moral rationality fails on its own terms as an account of rationality and, even when provided with Kantian underpinnings, has a serious rival as an explanation of the centrality of reason to morality. Its fault as a model of rationality is its assumption of the explanatory priority of the theoretical; its fault as an explanation of the centrality of reason to morality is that it leads to implausible accounts of moral development. Once again, there is a feminist connection. Feminist history of ideas suggests that, in morality especially, the explanatory priority of theoretical over practical rationality is historically conditioned and gender related. Abandonment of this assumption of priority as a heuristic in the project of understanding practical reason allows the dethronement of consistency as the measure of rationality. This makes salient different, hitherto unregarded features of practical reason and thereby creates a new problematic, a problematic that may reverse the familiar explanatory relation between the theoretical and the practical, replace unanswerable questions about morality with (at least prima facie) answerable ones, and offer a hope of collaboration rather than dissension over the answers both within and between philosophy, feminism, and cognitive science.

Objections to the N-D Model of Moral Rationality

The general objection is that the model as applied to morality fails to meet its own standards of rationality. On the N-D model, moral rationality takes two forms. At the lower level, it recognizes particular judgments as falling under general principles, an exercise of elementary deductive reason; at the higher level, it systematizes general principles, an exercise in axiomatic theory construction, a more complicated form of deductive reason. The measure of deductive reason is consistency: by deductive standards, judgment is rational only to the extent that it is consistent. But the consequences of applying this standard to practical morality reveal the impossibility of meeting it in this case. One could read this as showing that moral decision making is not at bottom essentially or entirely rational. Alternatively, it might show that consistency is not the right standard to apply, that moral rationality is being misconceived.

These consequences of making consistency the measure of practical moral rationality are as follows:

1. Particular moral judgments can be rational only as long as two or more general principles do not conflict over particular cases, for then conflicted particular judgments made according to either principle will be inconsistent with previous judgments made according to the other—unless, of course, principles accumulate ad hoc escape clauses. But then principles will be made according to judgments, not judgments according to principles. And unsystematized general principles typically do conflict over particular cases. Indeed, conflicts tend to multiply with time and the expansion of moral experience. Thus, if moral judgment remains at the lower level of the N-D model, it will tend to become progressively less rational as experience expands. Once conflict arises, systematization is essential if moral judgment is to remain rational.

2. Internal consistency may be preserved by systematization. But a set of conflict-generating general principles typically will be consistently systematizable in a number of different and incompatible ways. So what determines choice of systematization? Choice cannot be arbitrary, for that would make it irrational, and irrational choice of systematization would transmit irrationality to the judgments depending on the irrationally cho-

sen system. At this point, defenders of the N-D model usually switch from a purely formal, mathematical paradigm—axiomatic geometry—to one drawn from empirical science, usually physics (see Glover 1982). Applied to morality, the paradigm rules that the systematization must be chosen that best fits moral experience. Choice of systematization is rational in the same way as assessment of scientific theory: rationality is exercised in assessing the fit between moral theory and moral phenomena, and judgment of particular actions becomes a form of technology—the application of theory to practice. Judgments of theory fit may differ, but this is in the nature of moral experience, since in practice, rationally justifiable moral disagreement seems ineliminable (for in practice, moral principles conflict and can be systematized in different ways). Indeed, this is what makes free choice of theory—moral autonomy—indispensable. Rational moral autonomy is thus identified with rational assessment of theory fit.

This interpretation of rational autonomy is incoherent. It makes room for autonomous choice of theory only if different theories are equally defensible; if they are not, there can be no question of free choice because reason would oblige one to choose the best supported theory. But then autonomous choice of theory means choice on something other than the evidence, since by hypothesis all the candidate theories are equally defensible, that is, equally well supported by evidence. To concede the existence of ineliminable, rationally justified moral disagreement over theory on this ground is to concede that taken as a whole, the evidence fails to establish any one theory. In that case, autonomous choice of theory is not after all rational by the standards of the scientific paradigm. This undermines the conception of rational moral practice as technology, for a technology is only as reliable as the theory that backs it; to adopt a technology backed by a poorly established theory is (practically speaking) irrational, especially if the technology is applied when the theory is most doubtful, as is the case here: for the point of moral systematization was to deal with conflicted (that is, disputable) cases.

3. Practically oriented and theoretically uncommitted consumers of moral philosophy often express their uncommittedness by keeping all the theories in reserve and justifying particular judgments by reference sometimes to one theory, sometimes to another—on the apparently sensible ground that all the theories have something to be said for them. However, this too is irrational on the N-D model, since competing theories are constructed to exclude each other; the practice therefore amounts to entertaining multiple contradictions.

4. It is no solution, either, to explain differing judgments of theory fit as fitting the theories to different moral phenomena in the form of different moral intuitions, for this makes moral judgment ultimately subjective, and if it is so, what is the point of trying to make moral judgments rationally?

Rational methods are valued because they are more likely to produce right answers than irrational methods, but if morality is subjective, "right answers" can only mean "mutually consistent answers." But inconsistency is abhorred because of two inconsistent propositions, one must be false. However, if morality is subjective, a "true" moral judgment will mean one that accords with intuition, and it is certainly possible to have "inconsistent" intuitions. That is just what we have when moral principles conflict. In other words, the law of noncontradiction does not hold for moral intuitions. But then, what makes moral inconsistency rationally objectionable?

5. There is a further, ad hominem objection that can be made against those defenders of the N-D model who claim descent from Kant. As such, they presumably count themselves as moral rationalists on Kantian grounds. In that case, interpreting systematized principles, or moral theory, as confirmed and disconfirmed by moral intuition, is self-contradictory, for it combines a supposedly Kantian moral rationalism with an empiricist moral epistemology. Furthermore, Kant defined his project in moral philosophy by opposition to Hume; the two epitomize the polarization of eighteenth-century moral philosophy between reason and experience or feeling (passion). From this perspective the issue between them can be seen as one of moral psychology: Hume denied that reason unaided by passion could motivate action; Kant tried to show that it not only could, but must.

To do this Kant had to transform the traditional conception of practical reason. Nonmoral reasons explain action in terms of desire and the means to it. The end of nonmoral action is the satisfaction of desire, as the traditional conception says. Moral action, however, is not motivated by the satisfaction of desire. The structure of practical moral reason must therefore be different: it must be "pure"; that is, it must not depend on desire. The enemy in Kant's sights here is Hume's provocative remark in the Treatise that reason is and ought only to be the slave of the passions, and has no other office but to serve and obey them. Kant tries to save moral reason from this fate by reconceiving practical rationality in terms of the motivation of action not by desire but by recognition of submission to law. The analogy here is not with positive human law but with laws of nature, for the moral law, unlike positive law but like the laws of nature, is universal; it applies to all rational beings without exception.

Thus, if the N-D model accepts internal consistency combined with congruence with personal moral intuition as the mark of rationality in moral judgment, it traduces Kant. For this resiles from Kant's novel rationalist moral psychology in favor of a Humean one, since ultimately what explains or produces moral action is no longer reason but feeling, in the form of personal moral intuitions. What are these but Hume's sentiments of approbation and abhorrence in thin disguise? Consequently, reason becomes once more the slave of the passions and has indeed no other office but to serve and obey them, for its efforts are devoted to justifying whatever moral feelings one happens to have—and indeed, not even the complete set, but whatever subset avoids cognitive dissonance and can be most elegantly systematized (what else is "reflective equilibrium"?). "Rational" moral judgment thus becomes the product not of reasoning but of rationalization and a kind of intellectual aestheticism. Such a morality obviously cannot claim universality; at best, it can apply only to those who share one's moral feelings and intellectual predilections.

The N-D model of moral rationality thus fails dismally as an account of rationality, for what it calls rational moral judgment turns out, by its own, nomological-deductive standards, to be irrational—and by Kantian standards fails to be moral. It may be objected that to apply deductive standards as strictly as I have is to take the model too literally; moral reason is different—an analogue, not a species, of nomological-deductive reason. I would agree that moral reason is different; this is just what my objections suggest. But then the case for the nomological-deductive analogy needs to be made. Even if the N-D model is not taken literally, it remains unenlightening, for it raises, but is unable to answer, a whole series of questions that an account of moral rationality should answer— for instance, How is conflict of principles to be rationally resolved? What is rational choice between systematizations? How is rationally justified moral disagreement possible, and how can it be rationally resolved? What makes practical application of moral theory rational? Why should one try to make moral judgments rationally?

The N-D Model and the Centrality of Reason to Morality

Objection 5 above raised the question of the relation of the empiricist interpretation of the N-D model to Kant's account of moral rationality. Even if this objection succeeds—indeed even if all the other objections succeed as well—the model may still seem to have something right about it, and to be defensible on Kantian lines, as follows.

My objections appraise the N-D model as an account of what makes moral judgment rational from the point of view of moral success—as if reason were merely a content-neutral tool for getting out right answers to moral problems; and indeed, this is what the model itself suggests (since deduction aims at truth). But to judge reason by its success in producing truths or right answers is in fact to judge reason not by speculative standards, as I have been claiming, but by the standards of instrumental practical reason; this is what my question about the disvalue of inconsistency amounts to. But in the *Groundwork of the Metaphysics of Morals*, Kant showed that moral reason, though practical, could not be instrumental. Reason is something much more central to morality than a mere machine for producing right answers: it is constitutive of moral agency, and consistency is central to reason.

This objection points us to the heart of the matter. My answer to it divides into two parts, the first mainly concerned with the N-D model's relation to Kant, the second with my response to him.

First, I offer a preliminary ad hominem point about consistency: I would not deny that in what I said about consistency in objection 4 I am judging a speculative disvalue by practical standards. But if the alternative is to attribute an intrinsic disvalue to mere formal inconsistency that transfers to and explains moral value, it is a moot point whose side Kant himself would be on. Onora O'Neill has urged us not to separate Kant's overtly rationalist moral philosophy from the deeply antirationalist metaphysics of his *Critique of Pure Reason* (O'Neill 1989, chap. 2), where Kant insists on the impossibility of doing philosophy *more geometrico*: "In philosophy the geometrician can by his method build only so many houses of cards" (Kant 1933, A727/B/755; quoted in O'Neill 1989, p. 15).[1] This seems as applicable to moral reasoning on the N-D model as to Descartes. She also points out that in the *Critique of Practical Reason*, Kant asserts the primacy of the practical over the speculative: "Thus in the combination of pure speculative with pure practical reason in one cognition, the latter has primacy, provided that this combination is not contingent and arbitrary but a priori, based on reason itself and thus necessary.... Nor could we reverse the order and expect practical reason to submit to speculative reason, because every interest is ultimately practical, even that of speculative reason being only conditional and reaching perfection only in practical use" (Kant 1977, 120–122, quoted in O'Neill 1989, 3).

However, it may now seem that Kant's antirationalism can serve to weaken the force of my deductivist objections against the N-D model. Despite Kant's strictures against philosophy conducted *more geometrico*, he had nothing against the geometrical method as such; it just could not prove as much as Descartes thought it could. Similarly with moral reasoning *more geometrico*; my deductivist objections remind us of precisely this. The important thing is to avoid the temptation to take the empiricist turn in the (necessarily vain) search for moral certainty. Geometrical moral reasoning is not objectionable as long as it is aware of its limitations, as the recommended pluralist deployment of the N-D model clearly is. Willingness to resort to different theories in different circumstances need not be self-contradictory; it can be judicious asset stripping. Thus, Kant may put us on the right path even if his theory is objectionable. What is given primacy in the quotation from the *Critique of Practical Reason* is pure practical reason, and the supreme principle of pure practical reason is the categorical imperative, which we can accept without accepting all its theoretical underpinnings. If the categorical imperative is deployed as a check on the principles used in different systematizations, the N-D model can be saved; because of theoretical uncertainty, we will not have moral certainty, but we will nevertheless have moral objectivity, so will be able to tell acceptable systematizations from unacceptable ones.

Moreover, my question about consistency misses the point. It is true that consistency is a requirement of pure practical reason; judgments of what can and cannot be willed as a universal law are judgments of consistency. But this is not mere formal consistency, as is obvious from Kant's examples. The point of consistency is not to achieve moral certainty; to suppose that is to make a Cartesian mistake. Even alternative systematizations that pass the test of the categorical imperative may be possible, so that moral disagreement may remain ineliminable. The point is that since consistency is a practically rational requirement of the categorical imperative, reason is at the heart of morality. Thus, the N-D model, supplemented by the categorical imperative, is not primarily a machine for cranking out moral answers, but a development of this idea of the centrality of reason to morality; it is needed to cope with conflict of principles, which the categorical imperative does not address.

This brings us back to the heart of the matter, and me to my second and more radical response to this line of objection. I offer an alternative account of moral rationality to Kant's, but I am not going to lock horns with him in direct philosophical argument, and it is important to explain why.

The second part of my answer concerns the issue of consistency and the centrality of reason to morals, which raises knotty problems of Kantian exegesis. Nevertheless, it is not one I can deal with by engaging with Kant on his own terms, for one of the things at issue between us is where the burden of proof lies. (Fortunately, this means I can sidestep most of the problems of exegesis.) For (1) I agree with Kant that reason is central to morality, even that it is constitutive of moral agency; I deny that in order to be thus central and constitutive, reason must cease to be instrumental, that is, must be reconceived so as not to depend on desire. (2) It follows that Kant and I must disagree in what we mean by, what counts as, reason's being central to morality, or being constitutive of moral agency. (3) This raises a question about what motivates our respective accounts, what the accounts have to answer to; there would only be a point in trying to refute Kant directly if he and I saw the account of moral reason as having to answer to the same thing. But we do not, and cannot, for what I deny—that there is a fundamental distinction between moral and nonmoral reasons for action—is what his account has to answer to: he would regard it as an objection against his account if such a distinction did not emerge from it. Thus, individual statements from the two accounts cannot be compared directly; comparison of the accounts (for the purpose of choice between them) should take them as wholes and include comparison of their motivations. I shall attempt this in the final section of this chapter by trying to deconstruct Kant's moral psychology through a deconstruction of what motivates it.

Two Views of Moral Rationality

What, then, is my account? How can moral action be motivated by desire? The answer is, very simply, by being motivated by desire for the objectively good. Kant looks to moral rationalism for moral objectivity; I look to moral naturalism (though I shall not argue the case for it here; I have done so in Khin Zaw, in press). I am thus proposing a reversion to the traditional Aristotelian conception of practical reason developed by the Scholastics (most notably Aquinas) and deeply entrenched in Western culture.[2] (In fact Kant's attempt to reconceptualize practical reason as recognition of submission to universalizable law looks to me like an echo of the Scholastic conception of teleological natural law covering both the ethical and the physical; it is as if Kant were trying to reconceive this kind of explanation in terms of Newtonian science instead of Aristotelian science, that is, without teleology.) The differences between Kant's view and my version of the traditional view are set out in table 14.1. Since comparison of these two accounts of practical reason requires comparison of their motivations, my primary purpose here is not to mount a philosophical defence of my version of either. I shall not expound all the various items on the table in detail but explain and illustrate them as necessary.

I want first to relate the right-hand column to the earlier discussion of practical inconsistency, instrumentality, and desire. Note first the following points about "desire" and "good" as they there occur:

- "What is perceived as good is desired" is a synthetic a priori; the supreme principle of practical reason in V in the right-hand column expresses a kind of natural necessity. It is a necessary truth about the constitution of human nature.
- In "X is perceived as good," identicals cannot be substituted in X *salva veritate* (for example, if Millie is Joe's wife and the chair of the Women's Institute, "Joe perceives it as good that Millie be saved" does not entail either "Joe perceives it as good that his wife be saved," or "Joe perceives it as good that the chairwoman of the Women's Institute be saved").
- Joe may perceive the saving of Millie as good under one description of Millie, as objectively good under another, and as evil under a third.

Table 14.1
The Kantian and the traditional conception of moral reason

	Kant		Traditional View
I	Moral reason is practical but not instrumental	I	Moral reason is practical and instrumental
II	because moral action is not motivated by desire, but by reason	II	because moral action is motivated by desire for the objectively good, i.e., by virtue
III	so moral reasons are perceptions of pure reason (whose marks are universality and consistency), i.e., perceptions that a principle of action (=maxim/reason for acting) can be willed as a universal law.	III	so moral reasons are perceptions of the objectively good as such (and explanation of how we come by such perceptions can be empirical).
IV	Moral reasoning is reasoning about whether the principle on which one is about to act is a moral principle and consists of testing it by the categorical imperative.	IV	Moral reasoning is reasoning about how to realize the objectively good, i.e., about what to do for the objectively best.
V	The supreme principle of practical reasoning is the categorical imperative.	V	The supreme principle of practical reasoning is *Synderesis* (Aquinas)[b]: good is to be done and pursued and evil avoided.
VI	Rational agency depends on perceptions of reason (via the notion of freedom as self-legislating autonomy).	VI	Rational agency depends on perception of the object of action as good; moral agency depends on perception of the object of action as objectively good.
VII	The possibility of moral agency is a condition of rational agency.	VII	The possibility of rational agency is a condition of moral agency.
VIII	(Pure) practical reason is a condition of speculative reason ("the categorical imperative is the supreme principle of reason")[a]	VIII	Practical reason is an (empirical) condition of both the categorical imperative and speculative reason.

[a] O'Neill (1989, 1).
[b] Aquinas (1922, Ia IIae, q. 94) *Synderesis* is lucidly (and briefly) explained in Haldane (1989).

Thus, first, perception of/desire for the good, on this view, presupposes possession of concepts and distinguishes human from animal desire. Development of language enables finer-grained perceptions of the good. Second, desire is not a push but a pull. To conceive something as good (and hence desire it) is to have acquired an end in the sense of something one is disposed to go for (even if a very low-priority one). Human beings inevitably have multiple ends (simply *qua* animals with appetites and fears, as for food, safety, and stimulation), and multiple ends can get in each others' way on occasion (as with the fight-or-flight predicament in animals). The "reason" that distinguishes humans from other animals is the faculty that enables them to deal with such conflicts of ends in a different way from animals: instrumental practical reason.

Moral principles can thus be construed as moral ends (perceptions of objective good). As such, they, like ordinary ends, are liable to get in each others' way on occasion, but this is no longer a logical embarrassment, though it may sometimes prove severely taxing to practical reason. Nevertheless, the problem is amenable to solution by the same cognitive means as conflicts between nonmoral ends, for there too we want to have our cake and eat it, and the skills necessary to do so successfully will be useful in the moral domain. There they are even more necessary, for our perceptions are such that we feel we have no option but to find a way of having our cake and eating it—deciding to eat it instead of having it, or have it instead of eating it, simply will not do. (That is psychological, not philosophical, description.)

Now I think having our moral cake and eating it is not so hard as is usually believed; it looks hard only because the problem has been misconceived. The standard problem addressed by practical reason is not one of choice between two or more well-defined and incompatible alternatives, like an animal in the fight-or-flight predicament, for most of the time we are not in that situation. Usually we do not have to choose between already defined actions but can define our own, operating in an action-space that is constrained by moral and other perceptions but allows us some room for maneuver. The canonical form of the task for practical reason is thus to find (in the original sense of invent) an action that will nicely fit the space defined by the constraints. People do this all the time. For instance, judges when sentencing try to reconcile a number of conflicting goods, some of them objective (for instance, justice, mercy, the protection of the public, the reform of the offender or of public morals, exemplary punishment for the purposes of deterrence, education of or submission to public opinion). Ordinarily judges arrive at a sentence not by ruminating on the theoretical relationship between all these things but by hitting on a way of acting that will achieve all of the goods to some extent. In British courts this may, for instance, include stipulating or recommending the length of time to be served and delivering moral homilies from the bench while pronouncing sentence.

Getting this right—finding the action that best fits the space defined by the constraints, moral and otherwise—is clearly a complex, many-layered, high-level skill. Unfortunately we know hardly anything about what goes into it. One thing, though, is clear: since what is called for is invention, the process of finding the right action cannot be a deductive process, for deduction can only get out what has already been put in. Thus, theoretical knowledge may contribute to the process but cannot be its model. Learning from experience (whatever *that* is) obviously plays a part, as do sympathetic imagination and flexibility of mind. All these are aspects of human reason; it is the historical opposition of reason to feeling on the one hand, and to imagination on the other, that makes us think otherwise.

Since the deployment of these cognitive faculties takes place largely below the level of consciousness (and one can see good evolutionary reasons why it should), they all tend to get lumped together as a mysterious capacity called "judgment"—whose investigation should thus offer a rich and interesting field to cognitive science, once the philosophy has been properly sorted out. (I am not convinced it yet has.) Unfortunately again, understanding moral judgment has been bedeviled rather than helped by moral philosophers, who have delighted in tackling the problem by presenting each other with "fight-or-flight"–type moral examples with the object of putting pressure on each others' theories. Such examples generate the wrong questions, for the answers lead nowhere; the result is usually further baroque elaboration of an already minutely engineered theory, which may delight its maker and the fancy but does little to advance understanding and offers less to those still trying to cultivate virtue and live an objectively good life. Hard cases make bad law. Preoccupation with such examples has given the wrong focus to theory, making salient the exception and thereby distorting understanding of the rule.

Theories and Motivations

Let us return now to the question of what motivates the two accounts of practical reason. So far I have been stressing what distinguishes the traditional view of practical reason from the (Kantian) N-D model. I now want to draw attention to how much Kant's view and the traditional view have in common. Look again at table 14.1. In both columns, moral reason is practical, and it is essential to the determination of moral action; moral value is objective; and speculative reason depends on practical reason. Thus, both make practical reason central to morality. This suggests that to a large extent they are similarly motivated. What might this motivation be? In other words, what makes moral rationalism seem natural and plausible?

There have been some hints at an answer already. Western cultural and intellectual tradition makes reason central to human as distinct from animal nature, the key to both moral and cognitive maturation. Thus, in law, criminal responsibility was not attributed until "the age of reason"—in English law, originally decided by purely cognitive tests, such as the ability to do simple arithmetic, and set as low as age ten years. This suggests a web of background beliefs about reason, motivation, and the moral life enshrined in the Western cultural tradition that we may call traditional moral psychology. In this, reasons are not causes, so it is not to be confused with the "folk psychology" that some wish to eliminate from the philosophy of mind. The difference between reasons and rationalizations (which convinces some that reasons are causes) is not the difference between the efficacious and the inefficacious, but that between truthful and deceitful declarations of intent. Paul Churchland is right to want neuroscience to eliminate the psychology that makes reasons causes, but this is because that psychology is not really folk psychology at all. Folk psychology cannot be eliminated by cognitive science because it sets its agenda.[3]

Traditional moral psychology, I suggest, assumes (at least) the following:

1. Moral responsibility cannot be attributed before the development of specific powers of reason (though opinion on what these are has tended to vary).

2. Moral development ideally continues throughout life and consists in becoming wiser and better; it consists in advances in both moral understanding and virtue, which is the acquisition of a particular set of motivations. Moral understanding and virtue are interdependent and cannot develop without experience.

3. Moral understanding and virtue are developed by moral reflection on experience; moral reflection is an exercise of reason.

4. The beginning of moral maturity is deciding what to do for oneself, rather than obeying moral rules; deciding what to do for oneself requires moral reflection.

5. Moral deliberation is a further exercise of reason, called for only in difficult cases; their correct solution is its object, and mistaken solutions are possible. Capacity for effective deliberation is a mark of moral maturity, signaling the possession of wisdom (which few achieve).

6. Wisdom is the fruit of a well-conducted life, wide moral experience, and mature reflection and is not to be expected in the young or the vicious.

The traditional theory of moral rationality explains some of this traditional moral psychology. For example, an obvious candidate for the "necessary powers of reason" in 1 above is acquisition of a perception of the objectively good as such, or perhaps just a perception of something as objectively good (compare the traditional "knowing the difference between right and wrong," which used to be taken as an indication of criminal responsibility in English law); similarly, practice in moral reasoning as described in IV in table 14.1 would plausibly make one better at it, which could constitute becoming wiser and better; and II, III, and IV in the table suggest a path of lifelong development from amoral childhood to (let us hope) fullness of years and wisdom, which invites both philosophical and empirical amplification. I have not space to develop the theme here, but I suggest that this traditional moral psychology and traditional account of moral rationality jointly offer rich possibilities for cooperation in trying to understand human morality between, philosophy, developmental psychology, and neuroscience. The two accounts together suggest a whole host of questions inviting further philosophical and empirical investigation. Here are a few:

1. How do we come to acquire perceptions of something as objectively good, and of the objectively good as such? To what extent is it mediated by language?[4] What is this mediation? Why does moral responsibility depend on this (if it does)?

2. What is the "reason" that produces virtue, and how does it do it? Is this the same question as the previous one? Can psychology and philosophy help each other to interpret and develop Aristotle on this? (Consider, for instance, Aristotle's remark in *Nicomachean Ethics* II.5 that to be just and temperate, one must not merely do what the just and temperate man does, but do it *as* he does it. This has not impressed philosophers, but maybe it is as much practical advice as philosophical analysis. Compare traditional educational practice. Renaissance princes were urged to acquire virtue by modeling themselves on ancient exemplars, which one does by imagining one *is* them and seeing what they would do in one's situation—in other words, trying not just to do what they would do, but to do it *as* they would do it. This is simulation, which psychology is currently theorizing.)

3. What is "moral experience"? It does not seem to be just experience of morals. What does moral reason do with moral experience? (Here is a suggestion from maternal wisdom. In my youth, my mother once remarked that she did not want her children to live completely happy lives, as then they would not learn some of the things people should learn. They would lack essential moral experience. I think I now know what she meant. Compare what it is to spoil a child.)

4. What is the object of moral reflection (what exactly does it achieve)? What is its nature/structure (how does it achieve it)? What kind of relation is there between its methods (its content/structure) and its object (what it achieves)?
5. Substitute "deliberation" for "reflection" in the questions in 4.
6. What is wisdom?
7. How does instrumental practical reasoning do its job? (For instance, does it use cognitive gambits like the metaphorical transformation described in the next section, generalizable as pattern transference? Could connectionist models help here?)
8. Why is moral disagreement ineliminable? Is it something to do with the structure of the task of practical reason?

Now, how does the Kantian theory compare? How well does it explain traditional moral psychology? Before tackling this question, we need to consider a prior one: Does it have to? There is obviously a certain congruence between the left-hand column of table 14.1 and traditional moral psychology, suggesting some degree of traditional motivation; but I suggested earlier that part of the intention of Kantian theory was to institute a *new* moral psychology. To get some idea of what this might be, we need to apply it to an example. Here is a real one, which (by stipulation) exemplifies a morally worthy action.

A survivor of the Zeebrugge ferry disaster, invited to recall the experience in a radio interview some time after the event, was describing how he tried to get his wife and child out of the wreck (now on its side and largely submerged). As they groped through the darkness, they were joined by others, also looking for a way out. While advancing along the wall of a corridor—now their floor—they came to an open doorway, too wide to jump across, with corpse-laden water swirling below. Intent on saving his wife and child, the survivor felt around in the aperture until he found something that seemed to offer a firm footing; he then let himself down into the gap and stood on this base in such a way that his wife and child could use him as a bridge/stepping-stone, after which the others in the party followed, using him in the same way. At this point in his story, the survivor was interrupted by the interviewer with words to the effect: "Why did you let them do that? After all, your wife and child were safely across. Why stay for the others?" The survivor seemed somewhat nonplussed by this question. After a slight pause, he said with a half-laugh, "Well, sorry about the pun, but we were all in the same boat."

I propose the following as a Kantian account of this example: in saving his wife and child, the man was motivated by his desire that they survive: he wanted them across that gap. So acting as a bridge for them was prompted by desire, and therefore without moral worth (though in accordance with moral law). However, in remaining as a bridge for the others, he was not impelled by desire in the same way. Rather, he acted on the maxim, "When others are in need, help them," which he had previously freely adopted as a guide to practice because he perceived that he could will it as a universal law of nature; that is, that very perception motivated him, *qua* rational being, to act on the maxim. It had, so to speak, lodged the maxim in his mind as a potentially motivating force (a live-in, freely chosen lawgiver), so that when he saw others in need, he helped them. (His remark indicated his recognition that their need was the same as that of his wife and child.) Because of all this, his helping them has moral worth.

Note first that what the traditional theory of moral rationality represents as the difference between perceptions of subjective and objective good, both of which necessarily create desire for them, the Kantian account represents as the difference between desire and reason, both of which may cause action by determining the will. Thus, on the Kantian account, desire is a push, not a pull, as it is in the traditional account, where desire consists in being drawn toward an object seen as good. For Kant, in contrast, desire in the empirical realm of moral psychology becomes an irrational push, duly vetted moral principle a rational push, while intentional objects seem to be sidelined. The attractive power of the objectively good is recast as a form of rational compulsion: the categorical imperative determines the rational will as logic compels belief. The account in fact replaces a complex moral psychology whose central analogy is love, making action a reaching out toward something, and the intelligibility of action inseparable from the intelligibility of motivation, with an impoverished physicalist and coercive one in which action is conceived as motion imparted by psychological impulse, its intelligibility separated from the intelligibility of motivation: the ruling analogy is now Newtonian billiard balls, opening the way to eliminative materialism.

I have suggested that Kant rejected Hume's moral psychology, but now it appears that on a philosophically more fundamental level, he shares it. The philosophy of action at least looks the same, and perhaps the structure of the empirical moral psychology too. Kant could afford to concede this much to Hume, since for him, the empirical world is not where morality resides; for in the empirical world, we are not free, but determined.

If this is right, then Kohlbergian attempts at empirical investigation of Kantian moral development are absurd. Whatever they investigate, it is not Kantian moral development. It is unclear that there could even be such a thing, since Kantian moral worth seems to be an all-or-nothing, hit-and-miss affair: one either acts out of reverence for law, or one does not. Thus, the moral life becomes a matter of struggling to become (momentarily) a moral agent, perhaps scoring the occasional moral try, never knowing if one has succeeded, and with no obvious way of doing better. If this moral psychology explains anything in the empirical realm, it would appear to be a Lutheran conscience ever conscious of sin and unsure whether it is predestinately damned or one of the elect. But according to Kant, the home of morality is not there. Moral worth resides with reason, and it is beyond the reach of empirical investigation, since it consists not in mere cognitive capacity but in empirically inaccessible relations between cognition, will, and action. That this makes truly moral action in a sense unintelligible he concedes and accepts.

Morality and Gender

What motivates this radical retreat from empirical morality? Doubtless a large part of the answer lies in the history of ideas, philosophy's epistemological turn, and the special problematic created for Kant by his own critical philosophy. But I think there was also another, less well-recognized influence on his moral thought: eighteenth-century gender associations.[5]

I touched earlier on the polarization of eighteenth-century moral philosophy between reason and feeling. Many feminist writers (Lloyd 1984; Jordanova 1980) have noted related polarities, among them, humanity-animality, understanding-sensation, abstract-concrete, particular-general, maturity-immaturity, theoretical-practical—and, of course, masculine-feminine. Ludmilla Jordanova (1980) argues that such dichotomies are built into the structure of our thought, organizing it by a series of metaphorical transformations of one dichotomy into another:

> Our entire philosophical set describes natural and social phenomena in terms of oppositional characteristics. Each polarity has its own history, but it also develops related meanings to other dichotomies. For instance the pairs church and state, town and country also contain allusions to gender differences, and to nature and culture. Transformations between sets of dichotomies are performed all the time. Thus, man/woman is only one couple in a common matrix, and this reinforces the point that it cannot be seen as isolated or autonomous.[6]

Jordanova here seems to regard the gender dichotomy as on a level with all the others, but perhaps it is not, for why is the whole structure dichotomous at all? Perhaps because of the gravitational pull of that biologically almost universal dichotomy: sexual difference.

Jordanova goes on to trace the formative influence of these dichotomous metaphorical associations on eighteenth-century medical science, which was much concerned with sexual difference. She points out that softness (the key attribute for femininity) was applied both literally to the physical constitution of the female body to explain physical weakness, analogically to the female nervous system as a sign of typical feminine attributes like acute sensitivity and "irritability and suffering from vapours," and metaphorically as a symbol of intellectual inferiority and social dependence.

We find a cognate set of attributes assigned to women in an early work of Kant, *Observations on the Feeling of the Beautiful and the Sublime*. Rousseau, notoriously, promulgates differentiation by gender as a principle of education in *Emile*; Kant similarly distinguishes between male and female understanding along thoroughly conventional eighteenth-century lines:

> The fair sex has just as much understanding as the male, but it is a *beautiful understanding*, whereas ours should be a *deep understanding*, an expression that signifies identity with the sublime. To the beauty of all actions belongs above all the mark that they display facility, and appear to be accomplished without painful toil. On the other hand, strivings and surmounted difficulties arouse admiration and belong to the sublime.... The beautiful understanding selects for its objects everything closely related to the finer feeling, and relinquishes to the diligent, fundamental and deep understanding abstract speculations or branches of knowledge useful but dry. A woman will therefore learn no geometry.... In the opportunity that one wants to give to women to cultivate their beautiful nature, ... one will seek to broaden their total moral feeling and not their memory, and that of course not by universal rules but by some judgment on the conduct that they see about them.... Never a cold and speculative instruction but always feelings, and those indeed which remain as close as possible to the situation of the sex.... Nothing of duty, nothing of compulsion, nothing of obligation!... They do something only because it pleases them, and the art consists in making only that please them which is good.... I hardly believe that the fair sex is capable of principles, and hope by that not to offend, for these are also extremely rare in the male. But in place of it Providence has placed in their breast kind and benevolent sensations.[7]

Here Kant, following Rousseau and Burke, transfers to the moral and intellectual sphere the same dichotomies and the same metaphorical transformations that Jordanova finds in the physical sphere constructed by the medical scientists. "Beauty," for Kant, performs the role that "softness" performed in medicine. Rousseau also distinguished between male and female intellectual powers:

> The quest for abstract and speculative truths, principles, scientific axioms, everything pertaining to the generalization of ideas, is not within the competence of women, their studies should all be concerned with the practical; it is for them to apply the principles found

by man, and for them to make the observations which lead man to establish principles.... Woman, who is weak and sees nothing in the outside world, understands and estimates the motive powers which she can put to work to support her weakness, and these motive powers are the passions of man.[8]

Moreover, he thought morality should be gendered even in its content: rigorous honesty is to be cultivated in Emile, but girls should be encouraged to acquire sufficient cunning to evade rules they may not openly defy.[9] This seems to have suggested to Kant a way of achieving something perhaps even more radical by different means. Jean Grimshaw has pointed out that he assigns to men just those attributes that his later *Foundations of the Metaphysics of Morals* associates with true moral worth —reason, universal rules, principles—and to women, those without true moral worth—benevolent sensations, compassion, fine feelings.[10] However, I would put the point slightly differently: Kant's later work assigns to true moral worth the attributes that his earlier work assigned to men, and to absence of moral worth the attributes earlier assigned to women. The moral philosophy bends to the weight of a chain of symbolically related dichotomies.

Note too that female moral education proceeds along the lines recommended by the traditional account of practical moral rationality, according to which the art of instilling virtue consists, precisely, in "the art of making only that please...which is good." If this is Woman's version of moral principle, the dichotomies will remove Man's to the realm of deep, abstract, speculative, surmounted difficulties and painful toil—which is rather a good description of reading Kant on pure practical reason. In contrast, here is Rousseau's description of female practical reason:

> If you want to see someone in difficulties, put a man between two women with each of whom he has secret relations, and then observe what an idiot he looks. Put a woman in the same situation between two men, not an unusual event by any means; you will be amazed at the skill with which she puts both on the wrong scent, and gets each one to laugh at the other.... Thus each, happy with his lot, sees her always attending to him, while in reality she is attending only to herself... caprice would only repel, if it were not wisely managed; and it is in dispensing caprice with art that she makes of it stronger chains with which to bind her slaves. Whence all this art, if not from minute and continual observations which make her see what is happening in the heart of men at every instant, and which put her in a position to apply to every secret stirring she perceives the force necessary to suspend it or accelerate it? Now, is this art learnt? No;

women are born with it; they all have it, and men never have it to the same degree. Such is one of the distinctive characteristics of the sex. Presence of mind, penetration, minute observation, are the science of women; skill in making use of them, their talent.[11]

One cannot but sympathize with the desire—Kant's?—to distance moral reasoning from such goings on. But perhaps gender associations and an impoverished physicalist psychology concealed from him that instrumental reasoning has its glories too. Compare now another example. Here is Thucydides describing a supreme exponent of practical reason: the great Athenian general Themistocles:

Themistocles was a man who showed an unmistakable natural genius; in this respect he was quite exceptional, and beyond all others deserves our admiration. Without studying a subject in advance or deliberating over it later, but using simply the intelligence that was his by nature, he had the power to reach the right conclusion in matters that have to be settled on the spur of the moment and do not admit of long discussions, and in estimating what was likely to happen, his forecasts of the future were always more reliable than those of others. He could perfectly well explain any subject with which he was familiar, and even outside his own department he was still capable of giving an excellent opinion. He was particularly remarkable at looking into the future and seeing there the hidden possibilities for good or evil. To sum him up in a few words, it may be said that through force of genius and by rapidity of action this man was supreme at doing precisely the right thing at precisely the right moment.[12]

Notice how many attributes the great general shares with Rousseau's Parisian coquette. She too is supreme at doing precisely the right thing at precisely the right moment—the right thing to achieve her end, of course; but the same applies to him. And she does it on the spur of the moment, without studying the subject in advance or deliberating over it later, but using simply the intelligence that is hers by nature, since the matter she is concerned with (keeping both men happy here and now) certainly does not admit of long discussions. In order to succeed, her estimates of what is likely to happen if she does this or says that must be as reliable as his. No doubt she too could explain to her women friends the subject with which she is familiar—the passions of her men—and give an excellent opinion on the passions of theirs. (Note that Rousseau's point here seems still to apply if we generalize from the passions of particular men to human passions. Women have excelled as novelists; it is one of the few arts in which Western culture does not presume them inferior to men. Indeed it is almost regarded as their special property.) Conversely, if the general is to give the right orders in the thick of battle, he, like the coquette, must have presence of mind, penetration, and accurate observation, for he must continually see what is happening at the heart of the battle (which includes the hearts of the men fighting it), in a way that puts him in a position to apply to every incipient change he perceives the force necessary to suspend or accelerate it. Finally, both are described as achieving what they do by force of innate genius and rapidity of action.

If Rousseau was right to confer on women a special aptitude for instrumental practical reasoning, this, according to traditional moral psychology and traditional practical reasoning, would give them a better chance of achieving virtue and wisdom too. Could that be part of the reason why neither view has figured much in recent theory?

Acknowledgments

This chapter was written during a research fellowship at the Centre for Philosophy and Public Affairs at St. Andrews University. I am particularly grateful to the centre's director, Professor John Haldane, for extending the period of the fellowship to allow me to complete it, and to the Department of Moral Philosophy at St. Andrews for intellectual stimulation, encouragement, and generous hospitality.

Notes

1. The following discussion relies heavily on O'Neill, who makes the best case for Kant's moral philosophy I know.
2. I am grateful to John Haldane for pointing out to me the consonance between the view of moral reason I was developing and that of Aquinas. The latter view (with the essentials of its attendant philosophy of action) is referenced and accessibly summarized by him in Haldane (1989).
3. I have argued a similar point against rash claims made by Margaret Boden for cognitive biology in Khin Zaw (1980).
4. Perhaps Andy Clark's suggestions in chapter 6 about the place of language in a state-space morality might help here.
5. There is a more extended discussion of this topic in Khin Zaw (1992).
6. Jordanova (1980, 43).
7. Kant (1960, 78–81).
8. Rousseau (1951, bk. 5, 488; my translation).
9. Rousseau (1951, bk. 2, 93ff., 462–464).
10. Grimshaw (1986, 49).
11. Rousseau (1951, 485–486; my translation).
12. Thucydides (1979, 117).

References

Aquinas, St. Thomas. 1922. *Summa Theologica*. Translated by the Fathers of the English Dominican Province. London: Burns, Oates and Washbourne.

Beauchamp, T. L., and J. F. Childress. 1989. *Principles of Biomedical Ethics*. Oxford: Oxford University Press.

Cixous, H., and Catherine Clément. 1987. *The Newly Born Woman*. Manchester: Manchester University Press.

Glover, Jonathan. 1982. *Causing Death and Saving Lives*. Harmondsworth: Penguin Books.

Grimshaw, Jean. 1986. *Feminist Philosophers*. Brighton: Wheatsheaf.

Haldane, John. 1989. "Metaphysics in the Philosophy of Education." *Journal of Philosophy of Education* 23, no. 2.

Hume, David. 1978. *A Treatise of Human Nature*. Edited by L. A. Selby-Bigge. Oxford: Oxford University Press.

Jordanova, Ludmilla J. 1980. "Natural Facts." In *Nature, Culture and Gender*. Edited by C. MacCormack and M. Strathen. Cambridge: Cambridge University Press.

Kant, Immanuel. 1933. *Critique of Pure Reason*. Translated by Norman Kemp Smith. London: Macmillan.

Kant, Immanuel. 1953. *Groundwork of the Metaphysic of Morals*. Translated by H. J. Paton, as *The Moral Law*. London: Hutchinson.

Kant, Immanuel. 1960. *Observations on the Feeling of the Beautiful and the Sublime*. Translated by John T. Goldthwaite. Berkeley and Los Angeles: University of California Press.

Kant, Immanuel. 1977. *Critique of Practical Reason*. Translated by L. W. Beck. Indianapolis: Bobbs-Merrill.

Khin Zaw, Susan. 1980. "The Case for a Cognitive Biology." (Commentary). *Proceedings of the Aristotelian Society*, supplementary volume.

Khin Zaw, Susan. 1992. "Love, Reason and Persons." *Personalist Forum* 8, no 1.

Khin Zaw, Susan. In press. "Locke and Multiculturalism: Toleration, Relativism and Reason." In *Multiculturalism and Public Education: Theory, Policy, Critique*. Edited by A. Evenchik and R. Fullinwider. Cambridge University Press.

Lloyd, Genevieve. 1984. *The Man of Reason*. London: Methuen.

O'Neill, Onora. 1989. *Constructions of Reason*. Cambridge: Cambridge University Press.

Rousseau, Jean-Jacques. 1951. *Emile*. Paris: Classiques Garnier.

Thucydides. 1979. *The Peloponnesian War*. Translated by Rex Warner. Harmondsworth: Penguin Books.

Chapter 15

Moral Agency and Responsibility: Cautionary Tales from Biology

Helen E. Longino

I

Research on the biological dimensions of human behavior encompasses a broad variety of programs that incorporate vastly different conceptions of human nature as framing assumptions. Moving through literature on sex differences, aggression, crime, ethical behavior, and cognition reveals striking incongruities in conceptions of human behavior and in explanatory aims. In my own research as a philosopher of science, I have concentrated on understanding work in behavioral endocrinology and work in developmental neurophysiology. Although these areas are in many ways quite different from each other, they overlap in their common interest in characterizing the role of the brain in behavior. In behavioral endocrinology, much work through the mid- to late 1980s was characterized by what biologist Ruth Doell and I called a linear model. In this explanatory framework, which extends experimental research on animals to the human case, agency, intentionality, and responsibility are absorbed (and replaced) by hardwired neural organization. Among the many programs in developmental neurophysiology, I have been most interested in the theory of neuronal group selection. In this program, accounting for intentionality (or for a level of consciousness necessary for intentionality) is taken as a criterion of adequacy.

I shall present enough of these two research programs to show how they incorporate such diametrically opposed notions of agency and human personhood. My purpose ultimately is not just to cast doubt on the supposition that empirical science alone can decide such questions as the actuality of agency and intentionality, but also to suggest why moral philosophers might resist some forms of the rapprochement between cognitive science and moral theory that have been proposed to them. Like Paul Churchland, I think that neuroscience and cognitive science are not themselves the enemy of traditional moral theory and that the best protection against what might be perceived as unwelcome incursions is to know more about what neuroscience can show.

The two programs I shall discuss are not really comparable on a point-by-point basis. Both are in some sense about the role and function of the brain in behavior, and both are intended to illuminate facets of development. But behavioral endocrinology is interested in understanding the development of behavioral complexes, while developmental neurobiology is interested in the development of neural structure and function. The precise objects of research are therefore quite different: correlations between behavior, physiological status, and (primarily subcortical) brain anatomy in the one, and features of neural architecture, brain development, and cortical function in the other. The different questions researchers in these subfields bring to their subject matter carve out different domains within their area of common interest. Nevertheless, some comparative study is possible. I take it that a robust notion of moral agency involves something like decision making or intention formation that is based primarily on values, principles, desires, and beliefs that one endorses as one's own, and that is effective, that is, results in actions described by the content of the decision, or intention. The comparative question is, To what extent is such a notion tenable in these biological accounts of the role of the brain in behavior?[1]

II

The linear model in behavioral neuroendocrinology is a model of the transmission of the influence of prenatal physiological status via brain organization to postnatal behavior.[2] Researchers attempt to correlate prenatal or perinatal endocrine status with later behavior. They also correlate endocrine status with features of brain organization and attempt to establish experimentally a causal influence of the hormone exposures on the size of various features of the brain or on the distribution of hormone receptors in various regions of the brain. These observed effects of hormone exposure are taken as evidence that hormones organize the brain, either by establishing neuronal circuitry or setting neuronal threshhold levels for response to stimuli. The idea is that brain organization is the outcome of fixed and irreversible sequential processes initiated by pre- or perinatal hormone exposure (which in the ordinary case is itself an outcome of genetic structures). The organism is thus programmed to respond in a narrow range of ways to certain stimuli. Environmental, including social, factors can interact with brain organization to determine which of the behaviors in a given range will be produced (figure 15.1).

In a typical study, experimental mice treated with testosterone will attack a strange mouse introduced into the cage with a greater frequency than will mice that have either been gonadectomized prior to the critical

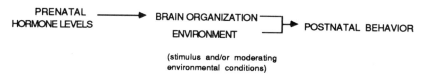

Figure 15.1
From Longino, *Science as Social Knowledge*. Reprinted with the permission of Princeton University Press.

period for organization or that do not secrete endogenous testosterone. Estrogen exposure at the critical period is similarly associated with the assumption of receptive sexual postures in the presence of other mice. These social behaviors are products of the interaction of hardwired brain organization and environment, the environment acting as both a stimulus and a modulator of behavior. There are many differences among species in the particular hormone-behavior productive sequences that can be established. The main point here is that behavior is treated as such a product in this approach, which integrates the classification methods of behaviorism with biology. When the theory is extended to humans, it offers a way of explaining behavior that does not require appeal to anything like the intentions or the self-consciousness of individuals. One area of human behavior to which it has been applied is presumed gender role behavior (figure 15.2).

The expression of gender role behavior is represented as part of a pattern of sex differentiation that begins with the chromosomes and extends to anatomy and physiology. Those differentiations include not just the structural elements directly involved in reproduction but also the brain and nervous system. That someone is aggressive or not, that someone is nurturant or not, that a child plays with trucks or with dolls is not a matter of how they think about themselves or of what they think about the situations in which these behaviors are elicited. It is not a matter of intention, that is, but a matter of physiological, anatomical, and environmental conditions over which they have no control.

This basic form of explanation is not limited to gender role behavior. Indeed in some ways gender or sex-differentiated behavior is just a pretext. The interest to biologists (and psychobiologists) of sex-differentiated behavioral systems is that they offer an opportunity to study bimodal behaviors that can be correlated with bimodal endocrine status. This makes them ideal as a model for studying the causal efficacy of hormones generally. Sex is both an accidental and an essential object of inquiry. It is accidental in that any kind of system displaying the same types of correlations would do; it is essential in that some such system is necessary,

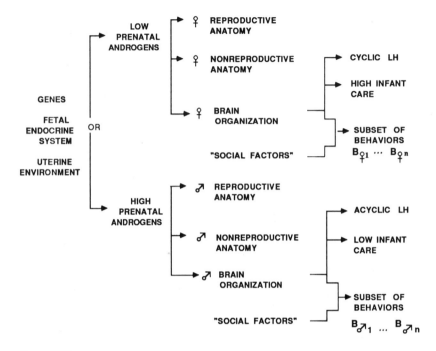

Figure 15.2
From Longino, *Science as Social Knowledge*. Reprinted with the permission of Princeton University Press.

and sexually dimorphic systems, especially their reproductive behaviors, provide the requisite features.

Although the linear model in behavioral endocrinology is most developed for sex-differentiated behavior, researchers have attempted extensions of the general form of the model to other areas, among them ethical behavior. Donald Pfaff, whose elegant work on the estrogen cycle in the rat remains a standard (Pfaff 1980), also ventured some opinions on the neurobiological origins of human values (Pfaff 1983). In this essay, Pfaff took reciprocal cooperation to be paradigmatic of ethical action. He argued that reciprocal cooperation could be analyzed into four modules, each of which he claimed could be accounted for in fairly straightforward neurophysiological terms: (1) representing an action, (2) remembering its consequences, (3) associating the consequences with oneself, and (4) evaluating the consequences. According to Pfaff, "Except where motor acts which require neocortex for their very execution are involved, ethical behavior may consist of a series of relatively primitive steps, in which, especially in their association with positive or negative reward, neurologically primitive tissues in the limbic system and brainstem, play the crucial roles" (Pfaff

1983, 149). Ethical behavior is thus analogous to reproductive behavior in being susceptible to biological analysis if properly decomposed into its constituent steps.

Pfaff's particular claims about the neurologic mechanisms underlying each module may well be contestable, but the importance of his account is the reconceptualization of action that it proposes.[3] An instance of reciprocal cooperation is the outcome of a sequence of physiological steps. Analyzed as behavior, it is a set of motions with consequences that stand in certain relations to the motions of other organisms. It is classifiable as reciprocal cooperation because it stands in those relations, and hence is describable by a rule of cooperation. It is not thus classifiable because the behaving individual has followed such a rule or intended to reciprocate some benefit or to elicit reciprocal behavior. In Pfaff's description, the idea of following a rule—the rule "do unto others as you would have done unto you" that he isolates as the ethical universal—is erased from the account. But a rule worth its salt as a rule, as Wittgenstein and others have taught us, is one that can be followed or not, that can be broken. There is no question here of the organism's deciding whether it will follow the rule, or of deciding which rules apply in a given situation, or of sorting through conflicting moral demands. If its behavior demonstrates the requisite relation to the behavior of other mice, a mouse can be as moral as a person.

Biologists are not alone in developing stories about the instinctual character of moral behavior. The social and behavioral scientists James Q. Wilson and the late Richard Herrnstein have, singly and jointly, integrated the results of a variety of social and biological studies to produce theories of both criminal and moral behavior. In *Crime and Human Nature* (Wilson and Herrnstein 1985), they argue for the influence of constitutional and genetic factors in criminal behavior. They cite data correlating body type, personality, and delinquency, along with data from twin and adoption studies suggesting a high heritability of criminality. They take the heritability to be genetic, not social. In addition, the fact that men commit the preponderance of violent crimes is also used to support the idea that there is something like a tendency to criminal behavior that has a biological basis.

Wilson pursues these notions further, not in relation to crime but in connection with what he calls "the moral sense," a set of innate dispositions to sympathy, duty, fairness, self-control (Wilson 1993). In the first book, there was a strong suggestion that the tendency to criminal behavior was in some way a physiological inability to control basic impulses. By contrast, the dispositions constitutive of the moral sense are based in a "prosocial" instinct that is the basis of this sense. Wilson hypothesizes that in the ordinary case they are capable of keeping basic self-seeking impulses in check because both originate in the same brain region, the limbic system (the same system Pfaff invokes). He cites in support of this notion Paul MacLean's hypothesis that the cingulate gyrus is involved in feeling states conducive to sociability (MacLean 1985).

Like Pfaff, Wilson is looking for a cause of what he takes to be moral behavior and specifically for a cause that bypasses the higher cognitive states based in the neocortex. Both Pfaff and Wilson develop explanatory schemata that reduce deliberation to relatively low-level processing, most of which can be carried out subcortically. It is thus subsumed within the paradigm of simple appetition realized in the limbic system. Ethical behavior loses its status as behavior chosen out of principle or in order to realize notions of right and good or other endorsed aims of the individual; that is, it loses its status as action. It becomes instead behavior that conforms to the author's or the author's culture's idea of what is right and good. The problem of understanding ethical action becomes the problem of producing desired behavior—of knowing what interventions are likely to increase its frequency. The implications for notions of agency and responsibility are clear. As traditionally understood, they no longer characterize human action. Decisions originate in the nervous system of the individual but not in those portions of the nervous system in which the higher cognitive processing involved in conscious inference, valuation, and deliberation is realized. Rather they originate in portions whose functional structure is determined in utero and remains fixed through an individual's maturity. Decision making is thus not subject to conscious deliberation and reflection. The effect of this shift is to reduce responsibility and agency to performance. The higher-level cognitive phenomena become at best epiphenomenal and at worst illusory relative to those neural processes that do cause behavior.

A strikingly different approach is to be found in another of the research programs currently pursued in developmental neurobiology. The theory of neuronal group selection (TNGS) is one of cortical development, developed specifically with a view to explaining or accounting for the high-level cognitive functioning that characterizes humans.[4] The theory involves both structural and functional aspects. The main structural claims are that cortical neurons are organized in groups (the cortical columns that have been identified by neuroanatomists), that the patterns of synaptic connectivity internal to these groups differ from group to group, that the groups are embedded in similar but nonidentical patterns of external, intergroup connectivity, and that the patterns of connectivity ensure that many groups will receive an identical signal from a given transmitting group. The differences in both internal and external connectivity among those groups mean that they will do different things with that signal.

Functionally, the theory postulates two fundamental processes: selection and reentrant signaling. Selection is the mechanism by which intergroup

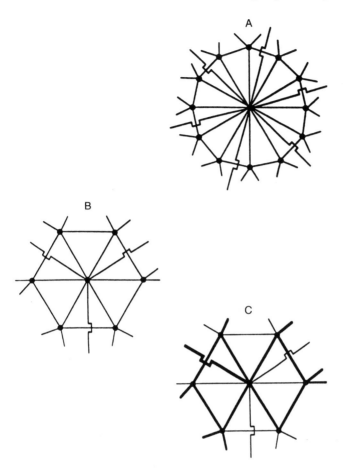

Figure 15.3
From Longino, *Science as Social Knowledge*. Reprinted with the permission of Princeton University Press.

connectivity develops. As the brain forms, the groups are densely but weakly interconnected. Functional patterns of external, intergroup connectivity will be selected from these preexistent connections. Figure 15.3 is a highly schematic representation of the stages of development of connectivity of one cell group with others.

The first stage (15.3a) represents the initial genetically determined system of cell groups, which is characterized by extensive redundancy both in connections and in number of cell groups. The second stage (15.3b) represents the formation of what Edelman has called primary repertoire. In this phase, certain synaptic connections are reinforced while others are idle as the nervous system gets "hooked up" by a process of trial and error and begins to function. Cell groups and synaptic connections that are not selected are eliminated from the system. The third stage (15.3c) represents the formation of functional connectivity, called secondary repertoire. This level of connectivity results from a second process of selection from primary repertoire, which occurs in the course of experience. This process consists in the amplification or inhibition of responses to a given signal pattern by a cell group rather than the elimination of synapses or cell groups. Secondary repertoire corresponds to the networks postulated by connectionist theorists in cognitive sciences, and the formation processes postulated are also similar.

Secondary repertoire varies in permanence. In humans, a fully functional nervous system is still developing during the first two years of life. Some of the stabilized connectivity becomes very stable through the reinforcement afforded by constant repetition, while some of it is in constant formation and re-formation in response to changing experience. The extensive redundancy that persists in secondary repertoire means that different combinations of cell groups will respond to the same signal type at different times in the same individual and that different cell groups and patterns of network activation will carry out the same recognition function in different individual brains. Edelman postulates in addition that after the formation of adequately functional secondary repertoire, there is still lots of primary repertoire left for additional development of secondary repertoire. This means, among other things, much flexibility and opportunity for additional learning. Because functional connectivity develops opportunistically, memory and other brain functions are conceived as distributed rather than localized or assigned to particular brain regions.

Consciousness or awareness is a function of what Edelman calls "phasic reentry." Each neuronal group is a processing unit that receives and transmits signals. Groups receive signals from noncortical areas, such as the thalamic nuclei, and from other cortical groups. Reentrant signaling occurs among groups, and among groups of groups (known as maps). The idea is that an original signal can be processed by a receiving group and reentered, or retransmitted to the original receiver, along with subsequent incoming signals. This process enables the association and horizontal integration of temporally successive signals and of multiple sensory modalities and the vertical integration of high-level (that is, more processed) signals with lower-level (or less processed) signals. This continuous multileveled reentrant signaling among neuronal groups makes possible categorization of features of one's environment, the distinction between self and nonself, and ultimately the emergence of higher-order consciousness.

Unlike the theories developed within the framework of the linear model, the TNGS is not presented as an explanation of specific behaviors, nor is such explanation an aim of the theory. The issue is not that of understanding the physiological conditions sufficient to produce the activities in question but of understanding in a general way what kind of neurophysiological processes are necessary for intelligent, reflective, self-conscious, creative activity. Nevertheless, it is possible to reflect on the form of explanation the TNGS makes possible. The theory is relevant primarily to nonreflex behavior—intentional behavior that involves the representation of self, action, world, and the consequences of action to oneself. Instances of such behavior would be understood to be an outcome of a complex set of neuronal interactions in the cortex. The theory postulates the continual alteration of neural networks in response to experience and action and a role for "self-inputs" (via the reentrant signaling mechanism) that include not just past associations (memory) of externally generated signals but representations of self as well, in the origination of action. Experience (including social experience) and self-image therefore are given primary roles within the framework of biological explanation of the behavior of species with a highly developed cortex.

Theorists focusing on higher brain function focus on human behaviors that as far as we know are unique to the species: those complex cognitive activities that include the construction of lengthy narratives in epic poems and novels, the composition of symphonies, the construction of undecidable mathematical theorems. We might include in this list engaging in moral reflection and critique. Because the seat of cognition and intention develops its unique or individually differentiated qualities in interaction with experience, multipotentiality, rather than limitation or determination, emerges as the character of the physiological contribution to behavior. This approach to understanding the role of the brain in mediating action returns both agency and responsibility to the person. The emphasis on the brain's plasticity and responsiveness to environment allows a role for processed social influence. To understand the brain as integrating physiological, environmental, memory, and self inputs, and then to understand decision and action as the result of that integration, rather than as the linear summation of physiological, environmental, and memory vectors, places control of action back in the individual without denying either the biological nature of consciousness or the role of social interactions in the formation of self.

III

What cautions do I wish to draw from these tales? Certainly not that the empirical sciences are irrelevant to moral theory. Where moral philosophers make empirical presuppositions, assumptions, or claims—for example, about the form of actual moral reasoning or about the process of development of moral concepts—the cognitive and neural sciences may well be informative or corrective in a way that makes a difference. The neuroscientist may challenge or expand our understanding of capacities, mechanisms, and habits of judgment. The examination of presuppositions is a reciprocal activity, however, and the philosopher has a continuing role to play in examining the philosophical presuppositions of the neuroscientist. I have suggested that as far as the linear hormonal model and the theory of neuronal group selection go, whatever degree of moral agency comes out is the degree that goes in. Radically contrary presuppositions, whether tacit or explicit, about the nature of human action have shaped the kinds of accounts generated; they have not emerged as conclusions from presuppositionless empirical research. Thus, my first caution is a warning against the application of conclusions in empirical science to philosophical problems without an adequate review of the diversity of research programs in a given field. The relations between philosophical inquiry and the sciences are too complex for uncritical appropriation of one by the other.

The second caution concerns the uses philosophers make of theories in the sciences whose generative questions are not only quite different from those that philosophers may have about a given domain but also differ within the sciences. These disparities in explanatory interests provoke some of the central tensions between traditional moral philosophers and scientifically oriented philosophers that have surfaced at this conference.

The linear theories in behavioral neuroendocrinology and the group selection theory in developmental neurophysiology have fundamentally different explanatory goals. In the linear model the object of explanation is the behavior or behavioral dispositions of individuals, and the aim is the development of linked causal laws asserting the dependence of particular behaviors (described as stereotypical motor outputs) on particular regions of the brain organized by genetically determined physiological events. In the case of presumed sex-differentiated behavior, the brain (in the ordinary case) registers genetically determined differential hormone exposure in its functional structure (first law), with the structure a vehicle for transmitting the influence of prenatal hormone levels to juvenile and adult behaviors (second law). When applied to the domain of moral behavior, this model is supposed to exhibit the dependence of particular behaviors that—for the proponents of the model—are identified as moral and immoral, on particular regions of the brain. These regions are organized to respond to certain stimuli in certain ways by genetically determined physiological events. The question the model is designed to answer is why one group of persons behaves morally while another does not. The moral philosopher seems to me perfectly within rights to complain that while the answer to this question is useful, indeed necessary, to the development of interventive techniques that might decrease the unwanted and increase the wanted behaviors, it does not explain morality. Nor does it take the place of systematic reflection on and critique of conceptions of the right and the good. To suppose the empirical sciences could do so is to commit the error of simple scientism.

The domain of morality is not just what people do but how they think about what they do, how it is that the behaviors the linear theorists are explaining come to be called moral or immoral, what that means to us, how having that designation or a cognate (good, virtuous, loyal) enters or should enter into a person's or group's decision to engage in or refrain from engaging in a particular course of action, and so on. And for some moral theorists it also includes the articulation of what moral action, given some specification of what "moral" means, consists in. These phenomena, which are part of the exercise of moral agency, involve high-level cognitive processing. Such processing is just the sort of thing the theory of neuronal group selection, or the family of theories of which it is a member, is trying to explain—not so much particular bits of cognitive processing but the capacity to engage in such processing at all. The explanatory goal of the theory of neuronal group selection is to answer the question, How must the biological substrates of capacities like the capacities constitutive of moral agency work? This biological theory leaves all the philosophical questions in place. It also leaves the causal question why some people behave in certain ways (for example, selfishly) and others in other ways (for example, unselfishly) in place. If humans are moral agents in a robust sense, then what the theory does say about the biological substrate implies that we cannot give a lawlike biological answer to that question, or at least not one that appeals to biochemical events in utero.[5]

Each theoretical approach formulates its research program in the light of some conception of human action, whether articulated or not. This conception permits certain sorts of question and forecloses others. Given the initial framing, studies and experiments can be designed to fill out the sketch. The way in which the linear model is expanded to the domain of the moral implies the irrelevance of philosophical inquiry to understanding morality, because it denies the presuppositions of the interpretive and analytic inquiry. The theory of neuronal group selection, by contrast, shares some of the presuppositions of the traditional philosophical approach, but it can be expected to provide a more nuanced understanding of the mechanisms of deliberation, intention, and action. It may change or expand the language in which we debate philosophical questions, but it does not moot them.[6]

The sciences offer a variety of ways of approaching a given subject matter, so we should be wary of staking our philosophical capital on the first biological theory that comes our way, since, especially at this early stage of development of the neurosciences, there is likely to be another theory around the corner adequate at least to erode our confidence in the first. To the extent that we put any weight on the "science" in "cognitive science," the same remarks hold. The best defense against scientism is not a rejection of science but more science.

Notes

1. The following section draws on and expands the discussion of these theories in Longino (1990).
2. Review articles, research reports, and hypotheses conforming to the linear model can be found in Goy and McEwen (1980) and DeVries et al. (1984). More recent work in behavioral neuroendocrinology is reviewed in the articles in Becker, Breedlove, and Crews (1992). The degree of conformity to the linear model varies among the authors represented in this collection.
3. The extent of the limbic system's role in memory is, for example, still under investigation and may not bear the weight Pfaff assigns to it.
4. The researchers most strongly identified with the theory are Jean-Pierre Changeux (1985) and Gerald Edelman (1987, 1989; Edelman and Mountcastle 1978). Connections to cognitive psychology are developed in some of the essays in Johnson (1993). The discussion here is based on Edelman's work.
5. Of course, the theory might be developed in such a way that it made sense to identify degrees of intentionality and to account for some, but not all, such differences in degree biologically. But this is a different explanatory problem.
6. The friendliness or unfriendliness of these theories to traditional philosophical concerns is not, of course, an argument for their truth or falsity. Part of my point here, however, is that the assessment of theories of such scope is more complicated than simple philosophical models of hypothesis testing suggest.

References

Becker, Jill, S. Marc Breedlove, and David Crews, eds. 1992. *Behavioral Endocrinology.* Cambridge, Mass.: MIT Press.

Changeux, Jean-Pierre. 1985. *Neuronal Man: The Biology of Mind.* New York: Pantheon.

DeVries, G. J., J. P. C. DeBruin, H. B. M. Uylings, and M. A. Corner, eds. 1984. *Progress in Brain Research*, vol. 61: *Sex Differences in the Brain.* Amsterdam: Elsevier Press.

Edelman, Gerald. 1987. *Neural Darwinism.* New York: Basic Books.

———. 1989. *The Remembered Present.* New York: Basic Books.

———, and Vernon Mountcastle. 1978. *The Mindful Brain.* Cambridge, Mass.: MIT Press.

Goy, Robert, and Bruce McEwen. 1980. *Sexual Differentiation of the Brain.* Cambridge, Mass.: MIT Press.

Johnson, Mark H., ed. 1993. *Brain Development and Cognition.* Oxford: Basil Blackwell.

Longino, Helen E. 1990. *Science as Social Knowledge.* Princeton: Princeton University Press.

MacLean, Paul D. 1985. "Brain Evolution Relating to Family, Play, and the Separation Call." *Archives of General Psychiatry* 42:405–417.

Pfaff, Donald. 1980. *Estrogens and Brain Function.* New York: Springer-Verlag.

———. 1983. "The Neurobiological Origins of Human Values." In *Ethical Questions in Brain and Behavior: Problems and Opportunities.* Edited by Donald Pfaff. New York: Springer-Verlag.

Wilson, James Q. 1993. *The Moral Sense.* New York: Free Press.

———, and Richard Herrnstein. 1985. *Crime and Human Nature.* New York: Simon and Schuster.

Chapter 16

Planning and Temptation

Michael E. Bratman

I

Much of our behavior is organized. It is organized over time within the life of the agent, and it is organized interpersonally. This morning I began gathering tools to fix the bicycles in the garage. I did this because I was planning to go on a bike trip tomorrow with my son and knew I needed first to fix the bikes. Just before I gathered the tools, I was on the telephone ordering tickets for a trip to Philadelphia next week. I did that because I was planning to meet my friend in Philadelphia then. If all goes well, each of these actions will be part of a distinct coordinated, organized sequence of actions. Each sequence will involve coordinated actions, both of mine and of others. These sequences will also need to be coordinated with each other. Such coordination of action—between different actions of the same agent and between the actions of different agents—is central to our lives.

How do we accomplish this organization? It seems plausible to suppose that part of the answer will appeal to commonsense ideas about planning. It is part of our commonsense conception of ourselves that we are planning agents (Bratman 1983, 1987). We achieve coordination—both intrapersonal and social—in part by making decisions concerning what to do in the further future. Given such decisions, we try to shape our actions in the nearer future in ways that fit with and support what it is we have decided to do in the later future. We do that, in large part, by planning.

I decide, for example, to go to Philadelphia next week. This gives me an intention to go, which then poses a problem for further planning: How am I to get there? My prior intention helps frame that further planning: it provides a test of relevance (only various ways of getting there are relevant options), and it provides a filter on options to be considered in my planning. It filters out, for example, an option of going instead to Chicago. Or at least this is how my intention functions if it is sufficently stable and is not reconsidered. In functioning in these ways, my prior intention helps explain why certain options get into my deliberation and others do not.

Why do we need stable intentions and plans to support coordination? Why don't we simply figure out, at each moment of action, what would then be best given our predictions about what we and others will do in the future if we act in certain ways in the present? One answer echoes the work of Herbert Simon (1983): we are agents with significant limits on the resources of time and attention we can reasonably devote to reasoning and calculation. Given these resource limits, a strategy of constantly starting from scratch—of never treating prior decisions as settling a practical question—would run into obvious difficulties. A second, related answer is that coordination requires predictability, and the actions of planning agents are more easily predicted.[1]

If we are to meet in Philadelphia next week, my friend needs to predict that I will be there then. The fact that I am planning to be there then helps support such a prediction. His prediction that I will be in Philadelphia next week need not depend on his detailed knowledge of my deepest values or on a prediction of complex calculations I will make just before leaving. He knows I am now planning to go, and that such a plan will normally control my conduct; in normal circumstances that will suffice to support his prediction.

I need not only coordinate with my friend and with my son; I also need to coordinate my own activities over time. Frequently this involves acting in certain ways now on the assumption that I will act in certain ways later. I would not bother with the tools in the garage if I were not fairly confident I would use them later. Just as my planning agency makes me more predictable to others, it also makes me more predictable to me, thereby supporting the coordination of my various activities over time.

I see such planning as a key to the phenomenon of intention; I call this the planning theory of intention.[2] Here I want to explore the relation between such views about intention and planning and some recent psychological theorizing about temptation—about the phenomena of giving into it at times and at other times overcoming it. I am interested, in particular, in the work of the psychiatrist George Ainslie, especially his recent and important book (Ainslie 1992). Ainslie seeks to describe what he calls a "mechanism for willpower" (144), and his discussion is fascinating and suggestive. But in the end, it does not take intentions and plans sufficiently seriously.

II

Suppose I am a pianist who plays nightly at a club. Each night before my performance, I eat dinner with a friend, one who fancies good wines. Each night my friend offers me a fine wine with dinner, and—as I also love good wine—each night I am tempted to drink it. But I know that when I drink alcohol, my piano playing afterward suffers. And when I reflect in a calm moment, it is clear to me that superior piano playing in my evening performance is more important to me than the pleasures of wine with dinner.[3] Indeed, each morning I reflect on the coming challenges of the day and have a clear preference for my turning down the wine. Yet early each evening when I am at dinner with my friend, I find myself inclined in the direction of the wine. If I were to go ahead and drink the wine, mine would be a case of giving into temptation.

Austin once warned us not to "collapse succumbing to temptation into losing control of ourselves" (1961, 146). Ainslie would agree. On his view, when I give into temptation, I do not lose control of my action. I control my action in accordance with my preference at the time of action; but this preference is itself at odds with central preferences of mine at different times. I proceed to sketch this story.

Begin with the idea that we frequently discount goods simply because they are in the future. Think of the utility of a certain good to me at a certain time as a measure of my preference for that good as compared with competitors. In temporal discounting, the utility to me today of a good that I would certainly get tomorrow is less than the utility it would have for me tomorrow, and this difference in utility is due solely to the temporal difference. Bill, for example, prefers cake now to ice cream now, yet he prefers ice cream now to cake tomorrow. If this shift in preference is not due to uncertainty about getting the cake tomorrow if he so chose, but is rather due solely to the different times at which Bill would get the desserts, then Bill's case is one of temporal discounting.

For such cases we can speak of the discount rate—the rate at which the utility of the future good is diminished solely by the perception that the good lies in the future. Perhaps such temporal discounting is irrational.[4] Still, many, including Ainslie, suppose it is pervasive. If so, we can expect certain cases of giving into temptation. If, for example, my discount rate is steep enough, the utility to me at dinner of my playing well that evening will be substantially reduced from its utility to me later that evening. The utility to me at dinner of drinking wine may then be above the utility to me at dinnertime of playing well later. If my action at dinner is determined by such preferences, I will drink the wine.

This story provides for some cases of giving into temptation but does not explain why earlier in the day I had a clear preference not for the wine but for the superior performance. According to Ainslie, if my discount rate for the wine is the same as my discount rate for the piano playing, if that discount rate is linear or exponential, and if at dinner I really do prefer to drink the wine, then I will prefer at breakfast that I drink wine at dinner. Merely by appeal to temporal discounting we do not yet account for the temporary reversal of my preference prior to dinner.

Ainslie argues that we can account for such preference reversals if the discount functions—representable by curves that map the utility to me of a certain event as a function of the time prior to that event—are not linear or exponential but, rather, sufficiently bowed so that the utility curves cross prior to the earlier event.[5] Suppose that my discount rate concerning wine is the same as that concerning piano playing,[6] and suppose that mine is a highly bowed discount function. (Ainslie's version is a hyperbolic discount function, grounded in Herrnstein's (1961) "matching law.") Then we can expect the following: From temporally far away, I prefer superior piano playing to wine at dinner. But at some time before dinner the utility curves may cross; there may be a reversal of preference. This reversal, however, is temporary; at some time after dinner, I again prefer superior piano playing to the wine.[7] If I have drunk the wine, I will then regret it. Ainslie claims that such highly bowed discount functions are pervasive and that they explain why temptation—understood as temporary preference reversal—is a common feature of our lives.[8]

Ainslie then asks how a rational agent can overcome the temptation predicted by such highly bowed discount functions. What rational strategies for resisting temptation are available?[9]

There are a number of ways I might try to respond to my knowledge this morning that at dinner my preference may change in the direction of the wine. I might, for example, make a side bet with you that I will turn down the wine, thereby generating new reasons for doing so, reasons that may block the reversal of preference. But there may also be available to me a way of thinking of my situation at dinner that will by itself enable me to overcome the temptation in favor of wine without a public side bet or its ilk.

When I compare at dinner (a) drinking the wine now with (b) not drinking now and, as a result, playing well later tonight, I prefer (a). But since I know I will be in a similar situation on many (let us say, thirty) future occasions, I can also compare the following sequences of actions, present and future: (c) my drinking the wine each of the next thirty occasions and (d) my refraining from drinking on each of those occasions.

On natural assumptions, at dinner on day 1 I will prefer (d)—the sequence of nondrinkings—over (c)—the sequence of drinkings; and I will have this preference even while, at dinner, I prefer (a)—drinking now—to (b)—not drinking now. This is because at dinner on day 1 I am still far enough away from dinner the next twenty-nine nights for my preferences concerning wine versus piano on those nights still to rank piano over wine. If I could choose the nondrinking *sequence*, (d), and if that choice would control my conduct, I could resist the temptation of wine each night and thereby achieve the preferred sequence.

Ainslie calls this approach to overcoming temptation the tactic of "personal rules."[10] This tactic depends on the agent's being able in some sense to "choose a whole series of rewards at once" (147). I agree with Ainslie that this is an important tactic. But to understand what this tactic involves, we need to ask what such a present choice or intention concerning future rewards or actions is.

The standard expected-utility model of rational action does not seem to have clear room for such a commitment to future action. On this model, an agent will have preferences, representable in terms of utilities, concerning present or future options or goods. An agent will have various expectations concerning the future—some conditional on present choices and some specifically concerning future preferences. An agent makes a choice about what to do now. There is, however, no further, distinctive state of the agent, which is her present choice or intention—and not merely her preference—concerning future action. Choice and intention are, strictly speaking, always of present conduct.[11] So we are faced with the question, What sense are we to make of the very idea of a choice or intention in favor of (d)?

One response would be to abandon the constraints of the standard model and to try to spell out how future-directed intentions function in a rational agent. That is the strategy of the planning theory of intention, and I will return to it. The more conservative strategy is to seek a substitute, while staying within the basic resources of the expected-utility model. This, I take it, is Ainslie's approach.[12] I proceed to sketch his version of conservatism and then to argue against it.

III

At dinner on day 1 I must choose between (a) drink wine at dinner tonight or (b) do not drink wine at dinner tonight, and as a result play better later tonight. My preference at dinner so far favors (a). But now I wonder what I will do the next twenty-nine nights: will I drink then, or not? When I try to answer this question, it is natural for me to see my present choice about whether to drink now as a "precedent": if I drink tonight, I will drink the other twenty-nine nights; and if not, not. That is, I believe (i) if I choose (a) then I will drink on each of the thirty nights, and (ii) if I choose (b) then on each of the thirty nights I will refrain from drinking.

My belief in (i) allows (a) to take on the expected-utility of (c), the sequence of thirty drinkings; and my belief in (ii) allows (b) to take on the expected utility of (d), a sequence of thirty nondrinkings. And my preference now concerning these sequences favors a sequence of nondrinkings, (d), over a sequence of drinkings, (c). So once I see things this way, I will prefer (b) over (a), and so refrain from drinking tonight. That, Ainslie supposes, is how I can use personal rules to overcome temptation. I "choose a whole series of rewards at once" in the sense that I choose a present option in part because of an expectation that if I choose that option I will choose a corresponding series of future options. My belief that (ii) "works by putting additional reward at stake in each individual choice" (Ainslie 1992, 153).

On this picture there is choice of present action, in part on the basis of beliefs that connect such choices with later choices of what will then be present actions. However, there is, strictly speaking, no choice or intention on day 1 to refrain from drinking on each of the next thirty nights. Or anyway, if there is such a choice, it is reduced to a choice of present action (namely, (b)) together with a *belief* connecting that choice to later choices (namely, a belief that (ii)). Such beliefs allow me to overcome my temptation to drink now. "The will," Ainslie says, "is created by the perception of impulse-related choices as precedents for similar choices in the future" (161). This story about overcoming temptation is conservative; it stays within the confines of the standard model. The choices and intentions that it countenances are strictly speaking, limited to present action. There can be a series of future choices—a series one can try to predict—but not strictly a choice of a future series.[13]

IV

I think that this account runs into a problem.

Consider my belief that if I choose (b) then I will abstain on each of the thirty nights, but if I instead choose (a) I will drink on all those nights. Is this a belief in the causal efficacy of a choice of (b)—that is, a belief that a choice of (b) would lead causally to later actions of nondrinking? Or is this, rather, only a belief that a choice of (b) would be good evidence that I am the sort of person—an abstainer—who will continue to choose nondrinking?[14]

Suppose my belief is only that a choice of (b) would be a predictor of analogous choices later.[15] Would this belief really give me reason to choose (b)? The problems here are familiar from discussions of Newcomb's problem.[16] On this interpretation of my belief I am thinking like a so-called one boxer if I reason from this belief to a choice in favor of (b). That is, in treating this belief as giving me reason to choose (b) I am choosing (b) not because this choice will have a desired result, but only because this choice is evidence of something else, something not affected by the choice itself. In choosing (b) I give myself evidence of, but do not cause, an underlying trait that will lead me to abstain later. There is a large literature that discusses whether such reasoning is legitimate (e.g., Campbell and

Sowden 1985). A common view, which I share but will not try to defend here, is that it is not legitimate. I will assume that, at the least, an account of rational "willpower" should not be forced to sanction such one-boxer reasoning. One-boxer reasoning is, at best, controversial in a way in which rational willpower should not be.

For Ainslie's approach to work, then, it should explain why I should see my choice to abstain on day 1 as a cause of later choices to abstain. Further, since we want a model of rational willpower, we need to explain how my choice to abstain on day 1 affects my reasons for choice on the later days, in a way that will lead me rationally to refrain on those days.

This is where the theme of "the strategic interaction of successive motivational states" is supposed to do work.[17] Ainslie wants to exploit a supposed analogy between a single-person case calling for willpower and a two-person repeated prisoner's dilemma, Ainslie writes: "A repeated prisoner's dilemma makes players predict each other's future moves on the basis of known past moves.... This is true whether the players on different days are two people, or a single person, ... as long as each plays repeatedly. Whether or not Monday's player is a different person from Tuesday's player, Monday's move will be the Tuesday player's best predictor of Wednesday's move and subsequent moves" (161).

Let us see how this analogy is supposed to work. Consider first a two-person case. Suppose Jones and Smith are in a situation of a repeated prisoner's dilemma in which each acts on alternate days. By performing the cooperative option on Monday, Jones may lead Smith reasonably to believe that he, Jones, will, at each of his turns, cooperate so long as Smith cooperates. Jones thus affects a belief of Smith's—a belief about how Jones would behave if Smith were to cooperate—and thereby gives Smith new reason for cooperating. And similarly for Smith, when Tuesday arrives. This may reasonably lead each to cooperate so long as the other cooperates on the previous day.[18]

Now return to me and the wine. Ainslie's idea is that by choosing not to drink on day 1 I give myself on day 2 evidence that if (but only if), on day 2, I do not drink then, as a result I will not drink on day 3. It is as if I, on day 1, am trying to engage myself, on day 2, in a mutually beneficial cooperative scheme. On day 1 I am like Jones on Monday; on day 2 I am like Smith on Tuesday. On day 1 I try to get me-on-day-2 to believe I am a kind of tit-for-tatter. My choice to abstain on day 1 causes me on day 2 (by giving me appropriate evidence) to believe that if I abstain on day 2, then, as a result, I will also abstain on day 3; and that belief helps provide practical reason for a choice to abstain on day 2. So my choice not to drink on day 1 is a cause of my rational choice on day 2, not mere evidence that I will make that choice on day 2. That, anyway, is the supposed analogy. It is in his use of this analogy that Ainslie sees himself as arguing that "the study of bargaining supplies the concepts needed for adapting behavioral psychology to apply to intraphysic conflict" (xiii).

Note that Jones and Smith are assumed to be rational agents. Their problem is how to cooperate given the reward structures and their rationality. Analogously, it is being assumed that on each day I am a rational agent. My problem is how, consistent with my rationality, I can overcome the temptation of temporary preference reversals. Ainslie's answer is that the solution in the intrapersonal case is analogous to a solution supposedly available in the interpersonal case when the "dilemma" is known to be a repeated one.

I agree with Ainslie that there are useful analogies between interpersonal strategic interaction and intrapersonal willpower. Jones and Smith each prefers the situation in which both cooperate over that in which neither does. Analogously, on each of the thirty days I prefer the situation in which I refrain from wine on all days to that in which I drink on all days. But I doubt that this analogy can do all the work Ainslie wants it to do. It is one thing to explain what is to be achieved by willpower, another to explain how it is rationally achieved. When it comes to explaining the latter, the analogy between one-person and two-person cases seems to break down. Or so I want now to argue.

It is important to be clear about what is needed here. It needs to be shown that my choice to abstain on day 1 is not merely evidence of an underlying tendency to make similar choices in the future but also a cause of such future choices. Ainslie tries to do this by seeing my choice on day 1 as evidence for, and so a cause of, a belief to be held on day 2. This belief on day 2 must satisfy two demands: it must be a belief that can function as part of a practical reason for a choice, on day 2, to abstain; and it must be a belief that is to some extent confirmed by the choice to abstain on day 1. The belief, on day 2, that may well be to some extent confirmed by the choice to abstain on day 1 is the belief that I tend to abstain. But that is not a belief that normally gives me, on day 2, practical reason to choose to abstain on day 2. The belief that if I had it on day 2 would give me reason to abstain on day 2, is the belief that if I abstain on day 2, then as a result I would continue to abstain. The problem is to say why that belief is confirmed by my choice on day 1.

Ainslie's purported solution to this problem depends on the analogy with the two-person case. On Monday Jones can, while performing the cooperative option, indicate that his continued cooperation is conditional on Smith's cooperation: he can indicate that he, Jones, is a conditional, not simply a nonconditional, cooperator. And even if Jones does not explicitly indicate this, it may be clear to Smith that this is true, given the payoff structure: for he and Smith are two different agents who, it is being assumed, have no intrinsic concern for each other; and Smith knows that

Jones awaits his, Smith's, choice. My choice on day 1 to abstain is supposed, by Ainslie, to be analogous to Jones's cooperative act on Monday. When I come to decide on day 2, I am supposed to be like Smith on Tuesday. I can see that my abstaining on day 1 indicates a willingness to continue to abstain so long as I abstain on day 2, just as Smith can see that Jones's cooperation on Monday indicates a willingness to continue to cooperate so long as he, Smith, cooperates on Tuesday.

But it seems to me that here the analogy breaks down. Suppose that on day 2, I recall having chosen to abstain on day 1. Why would I suppose that the person who made the choice on day 1 is waiting around to see if I will "reciprocate" on day 2 and will, only then, continue to "cooperate" on day 3? Unlike the two-person case, there is no other person. There is only me. Simply to suppose that I am waiting around to see what I do on day 2 before I decide what to do on day 3 is to beg the question at issue; for the question is why my earlier choice to abstain gives me reason for later choices to abstain. You can, if you want, talk of the earlier "person-stage" who decided on day 1 to abstain, and distinguish that stage from the present, day 2 stage. But that earlier person-stage is no longer around at all, let alone waiting to see what I do on day 2.[19] So we are still without an explanation of why I should reasonably see my present choice to abstain on day 1 as a cause of future, rational abstentions.

It might be replied that on day 2 I will know that if I do not abstain on that day, I will cease believing I am an abstainer. But even if this is true, it is not clear how it can help. A belief on day 2 that I am an abstainer does not give me practical reason for a choice on day 2 to abstain. That fact is not changed by the fact—if it is a fact—that if I choose on day 2 not to abstain, I will cease to have this belief. That the continuation of my belief that I am an abstainer is itself causally dependent on my choosing to abstain on day 2 does not show that my belief is really a belief that my abstaining later is causally dependent on my choice to abstain now. A belief that is itself causally dependent need not be a belief in causal dependence.

Consider a final reply: I know on day 1 that if I were to choose to drink the wine, then on day 2 I would believe I am not an abstainer. This would tend to make me, on day 2, discouraged about the possibility of my abstaining on day 2 and would thereby tend to make me more likely not to abstain. In that way I may see my choice on day 1 as causally influencing analogous choices later.[20]

The first point to note about this reply is that it does not involve the analogy with the two-person case. I am no longer seen as trying to convince my later self that I am a kind of tit-for-tatter. Rather, I am simply taking into account, on day 1, the psychological impact on myself of my choice. The second, related point is that the impact I am focusing on—

becoming pessimistic and discouraged—is not seen by me as giving me a new practical reason to drink on day 2.[21] This is a genuine disanalogy with the two-person case, in which information about earlier choices is supposed to provide a rational basis for later choices. Instead, what I am worrying about, on day 1, is that on day 2 I would be led to drink as a result of psychological causes that are not reasons for drinking. But the project is to explain how a rational agent overcomes temptations created by temporary reversals of preference.[22] That I have earlier failed to refrain from the wine is not normally a practical reason to fail again,[23] though it may be evidence that I will fail again. To see my earlier failure as ineluctably leading to my later failure is to be guilty of a kind of "bad faith" rather than to be functioning as a rational agent. So this last reply does not accomplish what we want.

So I am skeptical about Ainslie's "mechanism for willpower." The analogy between the intrapersonal and the interpersonal case does shed light on a parallel structure of payoffs. And we do sometimes treat our choices in some sense "as precedents for similar choices in the future" (Ainslie 1992, 161). But it is one thing to explain what is to be achieved by willpower, another to explain how it is rationally achieved. Absent an appeal to controversial forms of reasoning, Ainslie's model seems not to provide such a rational mechanism.

V

Return to the choice I face on day 1. Ainslie assumes that, strictly speaking, what I must choose between is (a) and (b). This assumption leads to a problematic story about a "mechanism for willpower." This suggests that we reconsider the assumption that the proper objects of choice and intention are limited to present options. Why not take more seriously the idea—supported by the planning theory of intention—that I really can choose between the sequences, or policies, cited in (c) and (d)? Given my clear preference for (d) over (c) I choose, and so come to intend, (d).[24] I make this choice believing that (iii) if I choose (c) then, as a result, I will drink on each of the thirty nights, and (iv) if I choose (d) then, as a result, on each of the thirty nights I will refrain from drinking. My intention in favor of (d) normally leads me rationally to abstain on day 1, and so on. That is how intentions normally function. This intention is a basic part of the relevant rational mechanism, a mechanism that is a part of our planning capacities.

Ainslie says: "The will is created by the perception of impulse-related choices as precedents for similar choices in the future" (161). On the alternative planning theory, the "will" is expressed also in choices of (and intentions concerning) future options, not merely in the perception of choices of present options as precedents for future choices. The primary basis for such choices and intentions is the desirability of those future options. As planning agents, choices and intentions concerning the future pervade our lives; they help structure the planning we depend on to achieve the benefits of coordination.

Ainslie and I agree on the importance of choosing a whole series of rewards at once. The issue is how to model such choices. I have considered three different models. On model 1, what is chosen is strictly a present option, but this choice is seen as evidence of a general tendency to choose similarly. This model sees a person with willpower as a one-boxer. On model 2, which I take to be Anslie's preferred model, I am still limited to choosing only present options. But that choice—for example, to abstain now—can be seen by me as causing later beliefs, which then ground later, similar choices. Finally, on model 3, the model suggested by the planning theory, I can make a present choice in favor of a valued sequence of future actions or a valued policy to act in certain ways on certain occasions. Such a choice issues in an intention concerning future conduct, and this intention is, normally, causally responsible for specific, future actions.

Once we take future-directed intentions seriously as basic elements of the psychology, we face a host of questions. We need a general account of how such intentions function in the practical reasoning of intelligent agents like us. We need an account of when an intelligent agent will settle, in deliberation, on certain intentions for the future, and of when she will later reconsider such prior intentions. This latter will be an account of the stability of intention (Bratman 1987, chap. 5; 1992). Part of an account of intention stability will focus on relevant mechanisms of salience and problem detection, for we do not want, as limited agents, constantly to be reflecting in a serious way on whether to reconsider our prior intentions. Finally, these accounts—of practical reasoning and of intention stability—will be linked: rationally to decide now, on the basis of present deliberation, to A later in circumstances in which I expect to have rational control of my action, I must not suppose that when the time and circumstance for A arrive I will, if rational, abandon that intention in favor of an intention to perform some alternative to A. (Call this the linking principle.)

Let us reflect briefly on the stability of intention. We are agents with important needs for coordination over time, both social and intrapersonal; and we are agents with limited powers of knowledge and limited resources we can devote to planning and reasoning. Given these needs and limits, what should be our strategies, habits, and/or mechanisms of reconsideration of prior plans?

We can begin by noting that there are criticizable extremes. On one extreme is the person who always seriously reconsiders his prior plans in the face of any new information, no matter how trivial. Such a person would constantly be starting from scratch and would be unlikely to achieve many of the benefits of planning. On the other extreme is the overly rigid planner—one who almost never reconsiders, even in the face of important new information. Most of us, of course, are somewhere in the middle between these two extremes, and that is in part why we are able to achieve the benefits of planning.

A theory of intention stability will need to say more about where in the middle instrumentally rational but limited planners like us should be. This is a big project. Without arguing in detail, however, I think we can reasonably make a conjecture concerning cases of temporary preference change. The conjecture is that instrumentally rational planners will not endorse a habit or strategy of always reconsidering a prior intention to A at t whenever they find themselves, at t, temporarily preferring an alternative to A. After all, such a habit or strategy would frequently undermine a valued kind of intrapersonal coordination, and it would do this in the service of a merely temporary preference. An instrumentally rational planner will have mechanisms and strategies of reconsideration that sometimes block reconsideration of a prior intention in the face of merely temporary preference change. (This is not to say that she will always resist all temporary temptations.)[25] This means that prior intentions and plans can have an independent role in rational motivation: once in place they can sometimes rationally control conduct even in the face of a temporary preference change to the contrary. (If, however, one does go ahead and act as planned, in the face of temporary preference change, there will also be a sense in which that is what one preferred to do. But that sense of "prefer" need not get at what explains one's action.)

A model of planning agency will differ in basic ways from a model of the agency of nonplanning agents—mice, perhaps. We may share with mice the generic characteristic of being purposive agents, and that may be an important commonality. But we are also a distinctive kind of purposive agent: we are planning agents. This is central to our ability to achieve complex forms of organization and coordination over time. It is a point familiar from discussions of free will that purposive agency does not ensure the kind of responsible agency we care about when we reflect on matters of moral praise and blame. The point here is a related one: purposive agency does not ensure planning agency.[26] We have as much reason to suppose that we have intentions and plans as we have to suppose that we have beliefs and desires; so we should resist the temptation to settle for a model of agency that ignores the difference between us and nonplanning, purposive agents.[27]

I have so far followed Ainslie's lead and focused on cases of temporary preference reversal.[28] But a full story about planning will also need to consider cases in which one's preferences and evaluations concerning one's options are instead coherent and stable. We can expect reasonable strategies of (non)reconsideration of prior intentions to respond differently in these different kinds of cases.

More generally, we can distinguish two aspects of reasonable intention stability (Bratman 1992; 1996; but see DeHelian and McClennen 1993). First, we want mechanisms and stategies of reconsideration that over the long run will suitably promote our efforts to achieve what we want and value, given our needs for coordination and our limits. But, second, if in a particular case, it is clear to the agent that, taking everything into account (including costs of reconsideration, replanning, reputation effects, and so on), she does better, in the light of her longstanding, coherent, and stable desires and values, by abandoning some prior intention, then that is what she should do. If appeal to the usefulness of some general rule of (non)reconsideration were allowed to block reconsideration even in such a case, we would be sanctioning a kind of planworship that would be unacceptable for a reflective agent.[29] The first aspect of intention stability helps explain why a reasonable planner may resist temporary preference changes of the sort that Ainslie emphasizes. The second aspect ensures that planners need not be overly rigid.

This supposes that longstanding, coherent, and stable desires and values may reasonably play a different role in rational reconsideration than merely temporary desires and preferences. Though I cannot argue for this here, it does seem to me a reasonable supposition, and one that Ainslie may want to share. For me to stick with my prior plan in the face of a clear, stable, longstanding, and coherent on-balance preference and evaluation to the contrary seems overly rigid. But if what conflicts with the plan is merely a temporary preference, the result of a temporary preference reversal, the charge of rigidity may lose its force, and a pragmatic argument in favor of a general strategy of sticking with one's plan in such a case may be cogent.

Rational nonreconsideration can, then, depend in part on the temporariness of a preference change. Consider, in contrast, Kavka's toxin case (Kavka 1983). A billionaire will give me a lot of money on Tuesday if I form the intention on Monday to drink a disgusting but nonlethal toxin on Wednesday. To get the money, I do not need to drink the toxin; I just need to intend, on Monday, to drink it. But I need to form this intention without exploiting any external mechanisms (such as a side bet) or forms of self-obfuscation. According to the linking principle, I can rationally form this intention only if I do not suppose that I will, if rational, abandon it when the time to drink arrives. However, it seems I would rationally abandon it; for on Wednesday I would have no good reason to drink the stuff.[30] Nevertheless, some philosophers have suggested that a pragmatic approach to rational reconsideration can justify a strategy that would lead to not reconsidering such an intention: such a strategy would, after all, help me to become rich.[31] But I think the second aspect of reasonable reconsideration will block this move. If I were to be faced with the toxin on Wednesday and were clear-headed about my situation, I would see that my longstanding, coherent, and stable ranking overwhelmingly supports my not drinking it; this is not merely a temporary preference reversal. So I should then reconsider and abandon a prior intention to drink, despite the attractions of a policy of not reconsidering in such cases. Knowing this, I will not be in a position on Monday rationally to decide to drink on Wednesday. So there are crucial differences between such cases and Ainslie's cases of temporary preference change. Planning agency, properly understood, provides a mechanism of willpower for cases of temporary preference reversal without a commitment to unacceptably rigid planning.

Acknowledgments

An earlier and shorter version of the discussion of Ainslie's book was presented at a symposium on his book at the Pacific Division of the American Philosophical Association, April 1994. There I benefited from discussion with Ainslie and the other contributors: Ronald De Sousa and Alfred Mele. Versions of this paper were presented at the April 1994 Conference on Mind and Morals and at the December 1994 Conference on Methods in Philosophy and the Sciences. Margaret Gilbert provided helpful comments on the latter occasion. I have benefited from the discussions at these conferences. I have also learned from written comments from Peter Godfrey-Smith and Alfred Mele, and discussion with Gilbert Harman, John Pollock, and Bryan Skyrms. My work on this chapter was supported in part by the Center for the Study of Language and Information.

Notes

1. Velleman (1989, 225ff.) makes a similar point.
2. I distinguish plans as mere recipes from plans in the sense of "planning to." I might have a plan in the first sense for cooking lamb and yet in no way intend to cook it. It is plans in the second sense that involve intentions: if I plan to do something, I intend to do it.
3. Ainslie would put this point by saying that the reward of better piano playing is larger than the reward of wine drinking, where reward tends to be understood hedonistically. But the central issues here do not depend on such a hedonistic approach, so I am trying to avoid it.

   Note also that what Ainslie would call the "reward" of better piano playing on any given evening is larger than that of drinking the wine at dinner, whether or not I resist the temptation to drink the wine on earlier days or on later days. Playing the piano well on a single, given night does not become a wasted effort if I play poorly on other nights. This feature of the example is built into most of Ainslie's discussion (see, for example, figures 3.3–3.5 in his chap. 3), and I will take it for granted in my discussion.

4. See Lewis (1946, 493), Rawls (1971, 420). See also Parfit (1984, 158ff.).

5. We can ask in what sense this is an explanation of such preference reversals, but I put such matters aside here.

6. A nonobvious assumption, but one that Ainslie makes throughout. However, as Green and Meyerson (1993, 40) write, "The exponential model may predict a preference reversal if the value of the discount rate parameter is inversely related to amount."

7. I am supposing that after dinner I can compare playing the piano well later with my having had the wine. I can ask which I now, at a time between the two, prefer. Ainslie does not talk this way, perhaps because he sees preference as tied to choices one can still make. But it seems to me we need something like this way of talking.

8. Indeed, Ainslie claims that an enormous range of phenomena can be seen as instances of preference reversal associated with hyperbolic discount functions. Some of the supposed applications of this model (e.g., to certain cases of psychological compulsion—see Ainslie 1992, 225–227) seem to me problematic. Austin claimed that Plato wrongly collapsed succumbing to temptation into losing control of oneself. I wonder if Ainslie sometimes wrongly collapses losing control into succumbing to temptation.

9. I think it is fair to put the question this way. Ainslie might object, however, that he is also interested in strategies that themselves involve forms of irrationality but still help us resist certain temptations. These would be cases of so-called rational irrationality. My discussion, however, seeks "mechanisms for willpower" that do not involve irrationality.

10. Though it seems to me more aptly called a tactic of "personal policies." See Bratman (1989).

11. This powerful idea is not limited to expected-utility theory. Consider the nineteenth-century John Austin: "It is clear that such expressions as 'determining,' 'resolving,' 'making up one's mind', can only apply in strictness to 'volitions': that is to say, to those desires which are instantly followed by their objects" (1873, 451).

12. See especially his central discussion on pp. 147–162.

13. This, anyway, is the most natural reading of most of what Ainslie says. Consider, for example: "But how does a person arrange to choose a whole series of rewards at once? In fact, he does not have to commit himself physically. The values of the alternative series of rewards...depend on his expectation of getting them. Assuming he is familiar with the expectable physical outcomes of his possible choices, the main element of uncertainty will be what he himself will actually choose. In situations where temporary preferences are likely, he is apt to be genuinely ignorant of what his own future choices will be. His best information is his knowledge of his past behavior under similar circumstances, with the most recent examples probably being the most informative" (147, 150).

    It seems clear here that Ainslie understands the choice of a series not as a present choice of future action or rewards but instead as a series of future choices, a series one may try to predict on the basis of one's knowledge of one's past choices. To "arrange to choose a whole series of rewards at once" is to arrange that there be a series of future choices.

14. In his recent discussion of Ainslie's views, Robert Nozick asks a similar question, but answers it differently (Nozick 1993, 19).

15. As Ainslie suggests at the top of p. 152. See also his remarks on p. 203, where he notes the relevance of Newcomb's problem.

16. This problem was originally described in Nozick (1969). This essay is reprinted, with a number of other useful essays on the subject, in Campbell and Sowden (1985). In the original example discussed by Nozick, a person, call her Sue, is faced with a choice between two boxes. The first box contains $1,000; the second box contains either

$1,000,000 or nothing. Sue can choose either just the second box or both boxes. Sue knows all of the following about her situation. First, there is an extremely reliable predictor who has predicted what choice Sue will make. Second, if the prediction is that she will choose only the second box, the predictor has put the money in that box. Third, if the prediction is that she will choose both boxes, the predictor has put no money in the second box. But, fourth, the predictor has already done whatever he is going to do: when it comes time for Sue to choose, the money is already there or it is not. A choice of only the second box is good evidence that the $1,000,000 is there; but it does not cause the money to be there. A person who supposes that the rational choice for Sue is to choose only the second box has been labeled a "one boxer". Nozick has recently returned to this problem in Nozick (1993, chap. 2). I will not discuss here Nozick's new approach to this problem.

17. Nozick notes that "doing the action now may have a minor effect on the probability of repetition in accordance with the psychologist's 'law of effect'" (1993, 19). But Ainslie's view needs a causal influence more significant than this. That is what the analogy with the interpersonal case is supposed to accomplish.

18. I put aside, in both this case and the single-person case, issues raised by the fact (if it is a fact) that it is known of a specific, last day that after that day there will be no further occasions for cooperation.

19. In this sentence I was helped by Döring (forthcoming), which makes a similar point in discussing work of E. F. McClennen.

20. Ainslie suggests this reply on pp. 150 and 203. See also Nozick (1993, 19). Ainslie, for example, writes: "Acts governed by willpower evidently are both diagnostic and causal. Drinking too much is diagnostic of a condition, alcoholism out of control, but it causes further uncontrolled drinking when the subject, using it to diagnose himself as out of control, is discouraged from trying to will sobriety" (203).

21. Alfred Mele helped me see that there may be cases in which this is not true. These will be cases in which the value to me of resisting a temptation depends on my resisting it most or all of the time. In Mele's example, the "reward" of not smoking on a particular occasion may depend on my generally not smoking; otherwise it may be a wasted effort. In such a case, if I am pessimistic about the likelihood of my resisting a certain temptation on later occasions, I may reasonably see my resisting this one time as a wasted effort. But, as noted above in note 3, this is not the structure of the cases Ainslie focuses on. In any case, for such cases we have no need for an elaborate story about the analogy with the interpersonal case.

22. Anyway, that is my project. Ainslie seems sometimes unclear whether it is his, as in the earlier quoted remarks from p. 203. (See above, note 20.) But this seems to me a conception of his project that coheres with his overarching idea that "the study of bargaining supplies the concepts needed for adapting behavioral psychology to apply to intrapsychic conflict" (xiii).

23. But see note 21.

24. There is an important complication. At dinner on day 1, after the preference reversal, I prefer the sequence consisting of drinking tonight and abstaining on the rest of the evenings (call it sequence (d')) to sequence (d). (See Ainslie's table 5.2 on p. 161.) However, prior to the preference reversal I prefer (d) to (d'). So, other things equal, which I choose will depend on when I make the choice. But once I do make the choice, I have a new intention or policy that may help shape later action. (I was helped here by discussion with William Talbot.)

25. A point David Gauthier helped me appreciate.

26. This raises the general issue of how to understand the relation between planning agency and morally responsible agency. Consideration of our planning capacities con-

trasts with consideration of distinctions (Frankfurt 1971) between first- and second-order desires, in emphasizing the temporal spread of our agency. So a focus on planning may be useful in understanding a relevant kind of unity of agency over time (Frankfurt makes a similar point in Frankfurt 1988, 175).

27. A temptation in the direction of, as we might say, a kind of genus-envy.

28. There are also cases in which the preference change is temporary if it is resisted, but permanent if it is not resisted. I might now decide not to accept a job offer I expect next week. But I might know that at the time of the offer, I will be tempted and experience a preference change. Will that change be temporary? It depends on what I do. If I resist and turn down the offer, I will soon be glad I did. But if I accept the offer, the effects of the job itself will be reinforcing in ways that will make me glad I accepted. So my preference change is, in such a case, only conditionally temporary—that is, temporary only if I resist. Some versions of the story of Ulysses and the Sirens also have this structure. I will not try to discuss such cases here.

29. This terminology aims to highlight the parallel with J. J. C. Smart's (1967) famous criticism of rule utilitarianism as potentially guilty of sanctioning "rule worship."

30. This assumes that on Monday my mere intention, even in the special, science-fiction circumstances of the toxin case, does not by itself amount to an assurance to the billionaire of a sort that induces an obligation to drink the toxin.

31. See Gauthier (1994, forthcoming), DeHelian and McClennen (1993), and McClennen (1990, chap. 13). What we say here may have implications for our understanding of rational, interpersonal cooperation; but I will not try to discuss these matters here.

References

Ainslie, George. 1992. *Picoeconomics: The Strategic Interaction of Successive Motivational States within the Person.* New York: Cambridge University Press.

Austin, John. 1873. *Lectures on Jurisprudence.* Vol. 1. 4th ed. London: John Murray.

Austin, J. L. 1961. "A Plea for Excuses." In J. L. Austin, *Philosophical Papers.* Edited by J. O. Urmson and G. J. Warnock. Oxford: Oxford University Press.

Bratman, Michael. 1983. "Taking Plans Seriously." *Social Theory and Practice* 9:271–287.

Bratman, Michael. 1987. *Intention, Plans, and Practical Reason.* Cambridge, Mass.: Harvard University Press.

Bratman, Michael. 1989. "Intention and Personal Policies." *Philosophical Perspectives* 3 (1989):443–469.

Bratman, Michael. 1992. "Planning and the Stability of Intention." *Minds and Machines* 2:1–16.

Bratman, Michael. 1996. "Following through with One's Plans: Reply to David Gauthier." In Peter Danielson, ed., *Modeling Rational and Moral Agents.* Oxford: Oxford University Press.

Campbell, Richmond, and Sowden, Lanning, eds. 1985. *Paradoxes of Rationality and Cooperation.* Vancouver: University of British Columbia Press.

DeHelian, Laura, and McClennen, Edward. 1993. "Planning and the Stability of Intention: A Comment." *Minds and Machines* 3.

Döring, Frank. Forthcoming. "Commentary on McClennen and Shafir and Tversky." In *Proceedings of the 1993 Cerisy Conference on "Limitations de la rationalité et constitution du collectif."*

Frankfurt, Harry. 1971. "Freedom of the Will and the Concept of a Person." *Journal of Philosophy* 68:5–20.

Frankfurt, Harry. 1988. "Identification and Wholeheartedness." In Harry Frankfurt, *The Importance of What We Care About*, pp. 159–176. New York: Cambridge University Press.

Gauthier, David. 1994. "Assure and Threaten." *Ethics* 104:690–721.

Gauthier, David. Forthcoming. "Intention and Deliberation." In Peter Danielson, ed., *Modeling Rational and Moral Agents*. Oxford: Oxford University Press.

Green, Leonard, and Myerson, Joel. 1993. "Alternative Frameworks for the Analysis of Self Control." *Behavior and Philosophy* 21:37÷47.

Herrnstein, Richard. 1961. "Relative and Absolute Strengths of Response as a Function of Frequency of Reinforcement," *Journal of the Experimental Analysis of Behavior* 4:267–272.

Kavka, Gregory. 1983. "The Toxin Puzzle." *Analysis* 43:33–36.

Lewis, C. I. 1946. *An Analysis of Knowledge and Valuation*. La Salle, Ill.: Open Court Publishing Co.

McClennen, E. F. 1990. *Rationality and Dynamic Choice: Foundational Explorations*. Cambridge: Cambridge University Press.

Nozick, Robert. 1969. "Newcomb's Problem and Two Principles of Choice." In Nicholas Rescher et al., eds., *Essays in Honor of Carl G. Hempel*. Dordrecht: Reidel.

Nozick, Robert. 1993. *The Nature of Rationality*. Princeton: Princeton University Press.

Parfit, Derek. 1984. *Reasons and Persons*. Oxford: Oxford University Press.

Rawls, John. 1971. *A Theory of Justice*. Cambridge, Mass.: Harvard University Press.

Simon, Herbert. 1983. *Reason in Human Affairs*. Stanford: Stanford University Press.

Smart, J. J. C. 1967. "Extreme and Restricted Utilitarianism." In Philippa Foot, ed., *Theories of Ethics*. Oxford: Oxford University Press.

Velleman, J. David. 1989. *Practical Reflection*. Princeton: Princeton University Press.

Index